Milk

Milk

The Biology of Lactation

Michael L. Power and Jay Schulkin

JOHNS HOPKINS UNIVERSITY PRESS   BALTIMORE

Johns Hopkins University Press
2715 North Charles Street
Baltimore, Maryland 21218-4363
www.press.jhu.edu

Library of Congress Cataloging-in-Publication Data

Names: Power, Michael L., author. | Schulkin, Jay, author.
Title: Milk : the biology of lactation / Michael L. Power, Jay Schulkin.
Description: Baltimore, Maryland : Johns Hopkins University Press,
    2016. | Includes bibliographical references and index.
Identifiers: LCCN 2015043849| ISBN 9781421420424 (hardback) |
    ISBN 9781421420431 (electronic) | ISBN 1421420422 (hardcover)
Subjects: LCSH: Lactation. | Breast milk. | Milk—Composition. |
    Milk—History. | Mammary glands. | BISAC: SCIENCE / Life Sciences /
    Biology / General. | SCIENCE / Life Sciences / Evolution. | MEDICAL /
    Nutrition. | SCIENCE / Life Sciences / Zoology / Mammals.
Classification: LCC QP246 .P69 2016 | DDC 612.6/64—dc23 LC record
    available at https://lccn.loc.gov/2015043849

A catalog record for this book is available from the British Library.

*Special discounts are available for bulk purchases of this book. For more
information, please contact Special Sales at 410-516-6936 or specialsales
@press.jhu.edu.*

Johns Hopkins University Press uses environmentally friendly book
materials, including recycled text paper that is composed of at least
30 percent post-consumer waste, whenever possible.

Contents

Preface

This book was both exciting and difficult to write, for the same reason. Exciting because knowledge of milk is expanding so rapidly, as is biological knowledge in general. Technological advances have enabled scientists to begin to unravel milk's biochemical complexity. Our understanding of the regulation of life's processes continues to progress, including a far better understanding of the many mechanisms by which the expression of genes is regulated. Our appreciation of the importance of the microbial communities that live on and in our bodies has increased many fold, as has our ability to investigate the composition and function of those microbiomes. It is an exciting time to be a biologist.

Of course all of this rapid advancement creates difficulties as well. Knowledge is changing so quickly that it is difficult to integrate and understand all the implications. This is true for all branches of biology; but the serious, in-depth study of milk is a relatively young science, and that contributes to both the excitement and the difficulty. The gentle river that was milk research is fast becoming a raging torrent. We are all too aware that our understanding of the field cannot be truly current, as we read as fast as we can merely to attempt to stay abreast of new research findings.

We hope we have produced a book that informs, stimulates, and even challenges thought about milk and lactation, its evolution, and its importance to modern human life. We certainly don't claim to have all the answers. Indeed, we don't even have all of the questions. But we believe that the questions we present about milk, and the eventual answers, are vitally important to understanding the evolution of mammals and the medically important concept of the developmental origins of health and disease in humans. We hope the readers enjoy the book, but even more, we hope it inspires scientists to further contribute to the study of milk.

Many researchers have informed and inspired us. We would like to acknowledge the scientific and collegial contributions of Lauren Newmark (nee Milligan), Katie Hinde, EA Quinn, and Robin Bernstein. They are among the new milk warriors that will carry the field forward. Olav Oftedal has contributed immensely to the understanding of lactation, and we are grateful for our time working with and learning from him. Much of our scientific viewpoint derives from the work of early students of physiological regulation, especially Curt Richter (1894–1988) and Eliot Stellar (1919–1993). Our understanding and appreciation of the complexities of the breastfeeding debates were enhanced by our collaborations with Jan Chapin and John Queenan.

Milk

Of Milk, Mothers, and Infants

The young mother hurries across the hot, dry sands, her four legs carrying her quickly toward her nest and the four precious eggs within, her tail swaying side-to side to aid her balance. The time is 280 million years ago, in the middle of the Permian period of Earth's history, a time when the continents of Laurasia and Gondwana were colliding and merging to form the supercontinent Pangaea. The warm, wet rainforests of the Carboniferous period were rapidly shrinking and even vanishing, and the landscape was changing to a more arid environment, dominated by conifers and including large swaths of desert. Life on land was becoming more difficult, as water became scarce. Hence the mother's hurry. She had needed to leave her nest to forage and find water, but that left her eggs vulnerable to dehydration in the hot, dry conditions. She could not leave them for long

She reaches her nest and all is well. No predators have found her eggs, and she was not gone so long that their water loss was critical. These eggs are not like the bird eggs of today or the dinosaur eggs of a mere hundred million years ago. They are not hard, calcified structures, but rather what are called parchment shell eggs, with pores that easily allow water to pass both into and out of the egg.

The mother eases herself onto her eggs and begins to gently rub her chest against them. This action stimulates a prolactin surge from her pituitary gland, which both calms the mother and stimulates the glands on her chest to begin to make a watery secretion. The physical contact with her eggs also causes a surge in oxytocin from her pituitary, which further stimulates these glands, ejecting the watery secretion onto her eggs. The fluid bathes the eggs, which begin to absorb the vital water, but also additional compounds, such as calcium, phosphate, sodium, and other minerals. Proteins in the fluid might also enter the egg, to be metabolized and used for growth, but perhaps also to perform important developmental functions, acting as growth factors and hormonal signals. Perhaps there are complex sugars (oligosaccharides) in the fluid. The fluid probably contains little if any lipid (fat), though the mother does secrete lipid from glands on her skin to aid her in combating water loss, so it is possible. There are also compounds in the fluid that are poisonous; not to the eggs and the developing embryos inside, but to bacteria, molds, and fungi. After all, the eggs and the mother's skin are now covered in a warm, moist solution, perfect for microbial growth that would be dangerous for the health of the mother and her eggs. One of these molecules likely was the bactericidal protein lysozyme, which today is found in many animal secretions, including mucus, tears, saliva, and, yes, milk.

For despite her reptilian looks, this young mother is a synapsid, an off-shoot of the vertebrate phylogenetic tree that diverged from the reptiles about 300 million years ago. Her descendants 80 to 100 million years later will be the stem mammals, the ancestors of all mammals alive now on Earth. In that sense she could be considered to be the mother of all mammals, and the watery secretion she deposited onto her eggs the rudimentary beginning of lactation and milk.

Fast-forward 280 million years and watch as a young mother is handed her newborn. She cradles it in her arms and brings its questing mouth to her breast. The neonate instinctively roots with its mouth, finds the nipple, and mother and child settle in for their first nursing session. The infant's suck-ling action stimulates the mother's nipples, sending nerve signals to her brain resulting in a surge of prolactin and oxytocin from her pituitary, stim-ulating milk production in the mammary gland, and initiating milk let down. Milk begins to be expressed from her nipples, eagerly consumed by the suckling babe. The milk flowing into the baby's mouth is a complex biochemical soup containing all that the baby needs to grow and develop. Nutrients of course; water, protein, sugar, fat, minerals, and vitamins, all the necessary ingredients for life are contained within this marvelous white fluid. But also so much more. Immunoglobulins that carry the mother's im-munization and disease history to her baby, enabling its immature immune system to combat infectious disease. Hormones and growth factors that regulate growth and development, signals the baby will eventually produce on its own, but supplemented and in some cases wholly provided by mom in early life. Compounds to combat bacterial and fungal infections of the in-testinal tract, such as lysozyme mentioned above. But also compounds such as oligosaccharides that encourage the growth of beneficial microbes in the gut, to produce the vitally important gut microbiome that contributes to our health and well-being. And newly discovered, a host of RNA molecules that have the potential to affect actual gene expression of the baby. Milk has come a long way since its ancient origins perhaps 300 million years ago. As our technological abilities to measure biological compounds increases, our understanding of the complexity and importance of milk continues to expand.

The time is ripe for a book that takes a broad comparative view of the evolution of lactation and how breastfeeding and mother's milk affect the health and development of infants. Biological science has progressed to where we are beginning to understand how epigenetic effects and complex regulatory signaling translate the blueprint provided by our DNA into the development of an independent, living being. The base of knowledge is finally sufficient to begin to examine milk from the perspective of regulatory and developmental biology; to look beyond its acknowledged nutritional importance and to investigate milk as a mechanism by which a mother profoundly influences her baby's growth, development, and future health.

Lactation Defines Mammals

The term "mammal" derives from the Latin word for teat (*mamma*). The mammary gland is tissue unique to mammals. All mammal mothers, from the egg-laying duck-billed platypus to human beings, feed their offspring with the glandular secretion called milk. Milk is the first food for all neonatal mammals, and for many it is the sole food for most of infancy. No matter what the adult diet—carnivorous, herbivorous, or omnivorous—all mammals start life as lactivores.

Lactation as a reproductive adaptation has fundamentally affected mammalian biology. It began the strong asymmetry in reproductive effort between female and male mammals, placing the main reproductive burden on mothers. It allowed the reduction of maternal resources deposited into the egg; unlike most other vertebrates, relatively little of the necessary resources for life are initially deposited into a mammalian egg. For monotremes (platypus and echidna) and the many species of marsupials almost all of the maternal transfer of resources for growth and development is accomplished through milk. In these species, milk supports offspring growth and development beginning from a fetal or even almost embryonic stage. Placental mammals also have extremely nutrient-poor eggs. When those eggs are fertilized they implant in the mother's uterus and develop a chorio-allantoic placenta, which nourishes the fetus by transferring nutrients and gases from maternal circulation to the fetus. For placental mammals, earliest growth and development is no longer dependent on milk, but after birth the now milk-borne transfer of maternal resources from mother to offspring continues.

Milk allows mammalian mothers to signal biochemically to their offspring over an extended period, guiding the development of their young. Species-appropriate nutrition is certainly delivered, but also far more. The list of bioactive molecules in milk continues to expand at a phenomenal rate as technology improves to enable milk to be fully analyzed. Immune factors, hormones, cytokines, enzymes, and even gene-regulating RNA molecules have joined the nutrients in the list of vital milk constituents. From implantation to weaning, mammalian mothers are "speaking" biochemically to their babies.

What Is Milk?

A basic biological definition of milk is a glandular secretion produced by mammalian mothers to feed to their offspring. This fairly simple definition encompasses a tremendous diversity of biological interactions. Milk is a biologically complex fluid that serves many adaptive functions. For neonatal mammals, mother's milk is the main if not sole source of nutrition for a substantial period of early life, and thus must provide the necessary nutrients for life, growth, and development. It is not surprising that milks of all mammalian species contain the same basic nutrients: water, sugars, fats, proteins, minerals, and vitamins. However, the relative proportions of these nutrients can vary quite substantially between species, and to a certain extent even within individuals of the same species and across lactation within the same individual. The milk of each mammal derives from a long evolutionary history and matches the birth condition and trajectory of growth and development for each species. Just as there is no one type of mammalian neonate, there is no one mammalian milk.

But milk is more than a food. Milk also serves to bolster and stimulate the neonate's immune system. Immunoglobulins, for example secretory IgA, IgG, and IgM, are expressed in high concentrations in mammalian milks, especially in the milk produced in the first few days after birth (colostrum). Milk is a mechanism by which the mother's immunological history can be transferred to her infant, to prime and stimulate the neonate's immune system as well as provide immediate immunological protection against diseases that the mother has experienced. Milk is an example of where the Lamarckian concept of acquired characteristics actually exists in nature. Much of an infant's immune function is initially provided by mom and reflects her disease history.

In addition, milk contains many maternally derived metabolic hormones (e.g., ghrelin, leptin, and adiponectin) and growth factors (e.g., epidermal growth factor [EGF] and transforming growth factor β [TGF-β]). These signaling molecules may have profound effects on fetal and infant development, from guts to brain. For example, human preterm infants are susceptible to necrotizing enterocolitis, a terrible inflammatory disease in which sections of the intestine die. In many cases the only treatment is to surgically remove those necrotic sections, in which case the individual will have a short intestinal tract for life. The mortality rate for this disease can be greater than 20%, and even in those infants who survive, the severe inflammatory response that occurs can damage other organs including the brain (Neu and Walker, 2011). Prevention is obviously greatly preferred, and human breast milk is one of the best prophylactics. Feeding human milk to preterm infants significantly reduces the incidence of this disease (Sullivan et al., 2010; Ben et al., 2012). The mechanisms behind the prophylactic properties of breast milk are unknown at this time, but may include the regulatory effects on gut development of multiple growth factors found in milk (Rautava et al., 2011), the effects milk has on the neonate's gut microbiome (Ganguli et al., 2013), or other factors that we have yet to discover. But something in mother's milk is good for baby guts.

The metabolic hormones in milk are thought to have profound effects on the development of neonatal metabolism. The hormones leptin and adiponectin found in milk are suspected of playing a role in the protective effect of breastfeeding against childhood obesity and later diabetes (Savino et al., 2009; Newburg et al., 2010). Adiponectin has strong positive effects on insulin sensitivity (Weyer et al., 2001; Cnop et al., 2003; Li et al., 2009; Ziemke and Mantzoros, 2010) and milk-borne adiponectin may play a role in shaping glucose metabolism in neonates. The ongoing human obesity epidemic and especially the troubling increase in obesity among children are strong confirmation that human metabolic phenotypes are plastic, and that they respond to environmental signals. Mother's milk (or the lack of it) provides one of the earliest signals. We will often refer to these types of molecules (metabolic hormones, growth factors, and even immune factors) as information molecules, as they function to transmit information between organs within the body. In the case of milk, they are capable of transmitting information between mother and child.

Many of these information molecules of maternal origin found in milk can also be found in amniotic fluid (Wagner, 2002; Wagner et al., 2008). The fetus swallows amniotic fluid, and thus is exposed to these maternal signals even before birth. In a sense, mother's milk is the continuation of signaling that used to pass through the placenta but was interrupted by birth. The mammalian lineage has a long evolutionary history of extended maternal care that includes the direct transfer of biological materials from mother to offspring. Evolutionarily, regulatory signaling via milk is older than the signaling via the maternal-placental-fetal connection. Lactation preceded live birth. Milk was the first mechanism for extended biochemical signaling from a mother to her offspring in the mammalian lineage. We argue that the extended regulatory influence a mammalian mother has over her offspring is a fundamental adaptation of mammalian biology, and it all started with milk (Power and Schulkin, 2013).

Philosophy of the Book

A book about the evolutionary history and significance of milk and lactation is an ambitious undertaking. We are not able to cover all aspects comprehensively; that would require a multivolume set, and even then it would be a challenge. Inevitably there will be areas of the topic that are not covered sufficiently or at all. For one thing, the field is changing quickly and knowledge about milk is rapidly increasing. Technological advances are allowing more milk constituents to be measured. Milk is a complex biochemical fluid; our techniques are just now beginning to allow us to explore that complexity in detail.

We have chosen to focus on the role of milk in guiding the growth and development of offspring, with an emphasis on the regulatory nature of milk and its links to the developmental origins of health and disease. Our thesis is that the evolution of lactation allowed maternal biochemical signaling to influence the growth and development of offspring across an extended period of time, a characteristic and unique feature of modern mammalian biology (Power and Schulkin, 2013). Thus, much of the book will involve exploring milk as an evolved maternal-offspring regulatory mechanism, often cooperative in nature, but on occasion a source of maternal-offspring conflict.

We focus on the milk of placental mammal species in this book primarily because those are the milks we have studied and thus where we have the most expertise and knowledge, but also because humans are placental mammals, and one of our goals is to illuminate the evolutionary history and underlying biology behind the role breastfeeding and breast milk plays in modern human health and well-being. We apologize to the marsupial lactation scientists for the placental mammal bias of the book. Marsupial lactation is fascinating, and arguably more complex than placental mammal lactation. It would almost have to be, considering the range of early developmental stages that are supported by milk in a marsupial, stages which are supported by the maternal-placental-fetal connection in placental mammals.

Science is an active, dynamic endeavor. Scientific knowledge continually changes as more data and evidence are generated. No matter how careful and thoughtful we are as scientists, time and the work of other scientists inevitably demonstrate that we got things wrong. Sometimes it is only the details that are corrected; but other times entire concepts and paradigms are overthrown. The painful beauty of science is that it is self-correcting. No matter how aesthetically pleasing an idea is, scientists reject it if the evidence doesn't support it. Rejection can be painful, and thus science takes a measure of courage. In certain sections of this book we will be anticipating, some might even suggest speculating, on where current knowledge will lead us in the future. Where we are inevitably shown to be wrong, we hope that at least we were productively wrong, and stimulated research and discovery.

Structure of the Book

The book is divided into four parts, with each part containing chapters organized around a theme. Each part is introduced by a brief overview of the chapters, outlining the primary concepts and ideas for the section.

The first part is organized around the comparative perspective and explores the evidence regarding the ancient evolution of lactation, its possible early adaptive functions, the novel molecules that arose in the lineage of animals linked to lactation and milk, and the co-opting of ancient regulatory molecules to serve new functions to support this novel reproductive adaptation. Part II examines the nutrient composition of milk; milk as a food. In these chapters we explore the variation in milk composition across and within species, the concept of a lactation strategy and how that affects milk composition, and how to compare the functional nutritional aspects of milks from different species that might, on the surface, seem quite different. In part III, we focus on the non-food aspects of milk; milk as a regulatory mechanism that allows mothers to influence the growth and development of their offspring. We briefly explore the concept of our microbiome; how we as individuals are fundamentally affected by the evolved microbial communities that coexist in us. Mother's milk plays important roles in establishing our microbiome, especially in the gut. Finally, we explore the concept of developmental origins of health and disease (DOHaD), and how mother's milk may influence later adult health and disease risks. In the last part of the book we focus on human milk, its particular novel features and why they may have evolved, and how breastfeeding affects the health of mother and child. We also include a section on our use of other species' milks, a uniquely human behavior.

One last comment on the structure of the book: We hope that readers will read and enjoy everything in the book, but we realize that our intended audience is quite broad and will likely vary in the level of detail they desire for many of the aspects of biology related to lactation. We have attempted to accommodate the expected variation in readership by employing boxes in each chapter where certain concepts and facts are discussed in greater detail, for those who are interested, without sacrificing the flow of the text. These boxes present additional content, hopefully of interest, but not required for understanding our primary messages regarding the evolution and function of that marvelous fluid, milk.

PART I

The Birth of Milk

Milk is of fundamental importance to every living mammal and every mammalian species has its own milk that has evolved under the unique selective and phylogenetic histories of each species. At the same time milk represents one of the oldest adaptations that arose in the lineage leading to mammals. Scholarly analysis of the developmental patterns of the mammary gland and of the molecular evolution of many of the novel molecules in milk indicates that lactation is ancient and that it has existed in some form for hundreds of millions of years, predating the origin of mammals by more than 100 million years (Oftedal, 2002a; 2002b; 2012). We are beginning to develop a good understanding of milks from modern mammals. The composition of milk from a large number of mammals has been assayed for its nutrient content and, to a lesser degree, other bioactive factors. Advances in technology allow scientists to explore milk in much greater detail, and the results confirm that milk is a remarkable and complex biochemical fluid. Our knowledge of the ancestral secretions from which milk arose are far less certain.

Although most of this book concerns modern milk, we can't ignore milk's ancient origins. The milk of today derives from ancient genetic and physiological adaptations that allowed a tetrapod ancestor of mammals some 250–300 million years ago (the synapsids) to secrete a fluid from a hypertrophied cutaneous gland. Whether we would consider that ancient fluid to be milk and those ancient female ancestors of ours to be lactating is an interesting question. We doubt that the fluid would have looked like any modern milk, and those ancient females probably were not nursing offspring (though it is possible, if the definition of nursing is broadened to include offspring licking a fluid from its mother's skin). But chemical analysis of that fluid would have found molecules that are also found in modern milk, and the function of the fluid was almost certainly related to protection and some form of provisioning of eggs/offspring (Oftedal 2002a; 2002b; 2012). It was also a successful adaptation. Only one lineage of those ancient synapsids survived through to the modern era: the lineage that evolved lactation. We propose that milk and lactation was an important adaptive characteristic that allowed these synapsid ancestors of ours to escape the Permian-Triassic extinction, survive the long era of dominance by the dinosaurs until their extinction at the end of the Cretaceous, finally to become, arguably, the dominant land vertebrates of today.

Many Animals Feed Their Offspring

We are comparative biologists, in the broadest sense. For that reason we have chosen to make the first chapter of our book an overview of how non-mammalian species have evolved similar, at least analogous, adaptations for transferring parental resources to their offspring. Milk is the uniquely mammalian way to feed babies; but it isn't the only way parents can provide their offspring with food. Many non-mammals also feed their offspring, and in some cases with substances derived from their own bodies. Although we believe the mammalian adaptation of milk has raised this concept to a fundamentally higher level, we suggest that the same parental regulatory control over offspring development may have evolved to some extent in many taxa. Indeed, further research may show that some taxa have matched (in an apples-and-oranges sense) what mammals do with milk.

In our brief overview we discuss examples of offspring-feeding adaptations from insects, spiders, fish, amphibians, and birds. In most cases, it is the mother doing the feeding, but in some species both parents are involved, and in at least one example it is the father who produces food for his offspring. Many of these examples are, in effect, forms of cannibalism. For example, in some spider species the spiderlings literally eat their mother. But there are examples of substances produced by the parent on which the offspring feed: eggs, skin, mucus secretions, and in some birds a secretion termed crop milk.

Relatively few of these instances of parents feeding offspring from their own tissue or secretions have been well studied with regard to the possible full range of parental resources that may be transferred. The assumption is that the prime purpose is to transfer nutrients, but in some species other resources (e.g., immunological molecules) have also been shown to be included. The evolution of a crop secretion (crop milk) in three lineages of birds (pigeons and doves, flamingos, and male Emperor penguins) perhaps comes closest to the mammalian adaptation of milk. In all cases parents certainly provide resources, in most cases nutritional; but in some instances, especially for crop milk, regulatory information may also be important.

Lactation Is Ancient

In chapter two we will travel far back in mammalian evolutionary history to before there were mammals, and to when vertebrates first began to live on land. Terrestrial living produced a novel set of adaptive pressures, many revolving around fluid balance. Water loss is a fundamental challenge to living on land. A crucial adaptation that enabled early vertebrates to invade land was the amniotic egg (see chapter 2; box 2.2). The rise of the amniotes (animals that produced amniotic eggs) began the large-scale invasion of land by vertebrates. More than 300 million years ago two lineages diverged from an amniote ancestor: the synapsids and the sauropsids. Mammals are the only living descendants of the synapsid lineage. All of the reptiles, including birds, are descendants of the sauropsid lineage. Lactation likely originated in an early synapsid ancestor of mammals, shortly after the synapsid-sauropsid split (Oftedal, 2012). In chapter two we review the evidence supporting an ancient origin for lactation (of some sort), as well as the theories regarding the original composition of the proto milk and its adaptive functions.

Novel Milk Molecules

Among living animals, milk is truly unique to mammals, containing many novel constituents. Since neither milk nor mammary glands fossilize, these novel molecules of milk are the best estimators we have of how long ago the constituents of milk (and thus lactation) came into being. There are proteins found in milk that are found nowhere else in nature: lactalbumin, caseins, and so forth. The evolutionary history of these proteins suggests they came into being as long as 300 million years ago, maybe more. Most derive from duplications of existing genes that subsequently diverged to attain new functions. Whatever their original function, they might have been important preadaptations that allowed lactation to come into being. There are even novel sugar molecules that exist only in mammals; e.g., lactose and oligosaccharides containing lactose. In chapter three we review the molecular evidence for the ancient origin of lactation, and discuss the functions, both modern and possibly ancient, of the novel molecules that arose at the beginning of lactation.

Ancient Molecules Co-opted to Have Lactation Function

Evolution works on existing materials with the occasional novelty produced by mutation. It can be argued that it is far more common that new adaptations arise from novel functions of existing features than from new, unique mutations. This seems particularly true of the functions of information molecules, which signal throughout the body, regulating and coordinating functions among end organ systems. Most signaling molecules have been shown to have multiple functions that vary according to the tissue in which they are expressed or on which they act, and by the biochemical context in which they are acting.

In chapter four we examine two molecules that are associated with lactation: oxytocin and prolactin. Both are ancient molecules with many other, well-established and important functions in the body. They also are logical candidates for a lactation-like adaptation. Oxytocin is linked with attachment behavior and with the process of birth/egg laying. Prolactin is associated with parental behavior, in both males and females, across many taxa. Indeed, prolactin is vital for the production of secretions fed to offspring from taxa as different as fish, birds, and, of course, mammals. We will briefly review the evolution of these molecules from ancient precursors and examine the range of functions of the descendant molecules in taxa from invertebrates to mammals.

Chapter 1

Feeding Offspring

The primary adaptive function of milk is to transfer resources from a mother to her offspring so that those offspring can grow and develop into independently feeding individuals. Milk is unique to mammals, but all animals have evolved mechanisms to transfer resources to their offspring. For some species it is all in the egg, with little or no investment after the egg is formed; a strategy truly fundamentally different from the mammalian strategy. But many species across a wide range of taxa provide parental care to their offspring. In some cases the care is as simple as protecting the eggs, either from predators, from dehydration, or by keeping them warm. In many species the care extends post-birth/hatching and can include feeding the offspring. Often the food is prey or other edible items the parents obtain and deliver to their offspring, but in some cases the nourishment offered is produced by the parents' body, or in extreme cases even is the body of a parent.

The diversity of evolved solutions to the common problem of providing offspring with the necessary resources to grow and develop is truly fascinating. We believe that it is instructive to consider these alternative adaptations for parental resource transfer in order to comprehend fully the implications for mammalian biology due to the unique evolved adaptation of lactation. In this chapter we briefly review parental feeding of offspring from substances created by the parent in species ranging from spiders to birds. We stress that this is a cursory review. We are not experts in any of these examples, and we have undoubtedly missed many amazing parental strategies. One aspect of parental care behavior that we will emphasize is whether the care could represent signaling from parent to offspring that might affect and even regulate development. For us, that is a key aspect of the importance of lactation and milk as a biological adaptation. Mammalian mothers are providing important, developmental signals to their babies over an extended period of time, providing the opportunity for controlled variation in phenotype. In the examples of parental care by other taxa given below we will examine the possibility that these non-mammalian adaptations can approach the level of parental control of offspring development that is the hallmark of mammals.

Parental Care

All parents provide their offspring with the biochemical necessities to start life. Not all species provide parental care. The term parental care applies to resources delivered to offspring beyond that of the original fertilized egg. Parental care takes diverse forms, from providing offspring (including eggs) with a safe, sheltered environment, to feeding them, to teaching them necessary skills for survival. A wide variety of taxa among vertebrates and invertebrates provides at least some form of parental care to their offspring. Maternal care is more common than paternal care, but both are widespread.

Parental care should be carefully distinguished from parental investment, as defined by Robert Trivers (1972; p. 139): "any investment by the parent in an individual offspring that increases the offspring's chance of surviving (and hence reproductive success) at the cost of the parent's ability to invest in other offspring." To be parental investment as Trivers defined it, care given to one offspring must reduce the potential to offer care to other offspring, either past (older siblings), present (current siblings), or future (reduced future parental reproductive success). Parental care by definition benefits the offspring, but it may or may not have a cost for the parent. For example, it is not clear that providing body heat by huddling with offspring in a nest necessarily imposes a cost on the parent. It may or may not. For example, it might depend on whether the parent is sacrificing the opportunity to perform other activities (probably not the case for birds during the night when they sleep, but possibly true during the day when they could be foraging), or on whether the heat trapped in the nest provides a challenge to the parent. For example, lactating rats nurse their pups in the nest longer when the cage temperature is lower (Croskerry et al., 1978; Leon et al., 1978). Nursing time appears to relate to maternal heat stress; the mother rat ceases to suckle her pups and leaves the nest when her body temperature begins to rise. The heat she provides to her pups in the nest may be valuable to them (certainly the milk is); but the rise in temperature of the nest may impose a cost on her.

But even though most examples of parental care likely do involve some cost to the parent that doesn't necessarily imply that the care represents parental investment, in the Triversonian sense. It may not affect the care given to other offspring, especially care given to future offspring. Although there may be some short-term cost to the parental behavior, not all costs carry over into the future. If the cost does not affect future reproduction by the parent, such costs likely do not represent parental investment (unless the care provided to one offspring decreases care given to older or same-age siblings; i.e., sibling competition/rivalry). In species that produce a single young with a significant amount of time between births (e.g., humans), Triversonian investment is primarily about reducing care that can be given to future offspring. The investment concept of Trivers differs subtly from the perhaps more common conception of a parent investing in children, in which the key aspect is the distribution of resources from the parent to the child (food, time, protection, and so forth) in the expectation that those transferred resources benefit the child (and thus, ultimately, the parent's reproductive success). Triversonian investment requires an additional factor; those transferred resources must in some way decrease the ability of the parent to provide resources to future offspring, and thus decrease the parent's future reproductive success. There are many examples where parental behavior clearly represents investment. A parent interposing between its offspring and a predator, risking injury and even death, certainly represents potential parental investment. However, there are also plenty of examples of parental behaviors that impose a cost, but that cost may be transient, and not have any reliable effects on future reproduction. Those instances of parental care do not represent parental investment.

For example, all lactating mammals lose bone mineral. Women can lose anywhere from 3 to 10% of their bone mineral while breastfeeding (Prentice, 2000). This bone mineral loss can be considered a necessary cost of successful reproduction, as it transfers maternal mineral stores (mostly calcium and phosphate) to offspring to build bone; but does it represent parental investment? The bone mineral is rapidly regained post weaning (Kalkwarf et al., 1997; Prentice, 2000) even if a second pregnancy occurs within a year postpartum (Sowers et al., 1995; Laskey and Prentice, 1997). Premenopausal women who have breastfed an infant are not at risk for lower bone mineral density compared to women who have never breastfed (Canal-Macias et al., 2013; Tsvetov et al., 2014). In postmenopausal women there is some evidence that prolonged breastfeeding is associated with lower bone mineral density (Okyay et al., 2013; Tsvetov et al., 2014), though other studies have found either no effect or a positive effect from having breastfed (Kritz-Silverstein et al., 1992; Melton et al., 1993). Interestingly, in the study by Okyay and colleagues (2013) parity had a protective effect; breastfeeding more babies reduced the risk of low bone mineral density.

Under conditions of vitamin D and calcium sufficiency, the bone mineral loss during lactation appears to be a transient phenomenon, without negative future consequences. However, under conditions of prior deficiency in either vitamin D or calcium, lactating women (or other female mammals) are at greater risk of bone fractures, which could lead to morbidity and mortality, reducing future reproductive success. So does this mean that bone mineral loss during lactation is maternal investment for non-healthy females, but not for healthy females? Distinguishing between the costs of parental behavior, the concept of parents "investing" in their offspring analogous to investing money (allocating now in hopes of future returns), and Triversonian parental investment (allocating now at the cost of being less able to allocate in the future) can be difficult.

Protection

Providing a protected environment is probably the most common form of parental care. Female crocodiles spend time near their nests and will attack and drive away potential egg predators. There are species of spiders in which the female guards the egg sac and maybe the spiderlings for some time. Many of these spider species are communal/social species, and females may even watch over spiderlings that are not her own. In this case care consists of protection of vulnerable young, which could have a cost but does not represent a direct transfer of resources.

Some frog species carry their tadpoles on their backs; sometimes it is the female but in other species it is the male. The female gastric brooding frog, unfortunately probably now extinct due to habitat loss, went even further. She swallowed her fertilized eggs and the tadpoles developed in her stomach. The female's stomach was transformed from a digestive organ into a brooding pouch, in part due to prostaglandins excreted by the eggs and tadpoles. She quit feeding for two weeks as the eggs hatched and the tadpoles developed into froglets, at which point she regurgitated them into the world. There is no evidence that the female provided anything other than a protected environment for her tadpoles and a very large yolk sac to supply the resources to develop within that environment. All the biochemical resources and signaling were provided with the formation of the eggs, and the parental care represents provision of a safe environment.

Male seahorses and pipefish carry the eggs they have fertilized in a pouch. In the popular literature this is often stated as male seahorses getting pregnant and giving birth. A fair enough analogy, in a loose way, though there are fundamental differences between this type of "pregnancy" and pregnancy in species with internal fertilization and embryo retention within the body. The seahorse pouch is external, and skin separates the embryos from the male's immune system. Still, this unique adaptation not only protects the developing seahorse embryos from predators, but for at least some species the fluid inside the pouch is maintained at a different osmotic balance than that of seawater. The pouch fluid changes over time to match seawater more closely as the embryos develop and approach the time when they will be "born" (Partridge et al., 2007). A study has shown that seahorse embryos develop better with lower mortality in this modified fluid in comparison with embryos developing in seawater (Azzarello, 1991). Whether this parental behavior represents only protection from the external environment (e.g., from the potential desiccation due to the salinity of seawater), or whether there may be some important developmental signaling occurring between father and offspring is not known. In a study of the broad-nosed pipefish, embryos were radioactively labeled and parental uptake of the label was demonstrated, showing that material certainly passes from embryo to father (Sagebakken et al., 2010). This probably represented some form of reabsorption of failed embryos post mortem; but it is possible that secretions from the pouch either nourish and/or provide biochemical signals to the embryos. Seahorse fathers may provide their embryos with more than just protection.

Birds provide heat for their eggs as well as protection. Heat can be a signal and a necessity. In many species, males and females take turns sitting on the eggs, transferring body heat to their developing offspring. The Australian brush turkey provides heat for eggs in an unusual way. The male builds a large compost pile and attempts to attract females to mate with him and lay their eggs in his pile. The eggs are warmed by the heat of decomposition of the compost. The male will stick his beak in the pile and rearrange the pile and make other adjustments to maintain an appropriate pile (nest) temperature. The sex ratio of the hatchlings from the compost-pile nest depends on the temperature of the pile, though this is explained by differential mortality between the sexes at different temperatures, not some form of temperature-dependent sex determination (Göth and Booth, 2005). But certainly the amount and consistency of heat provided to eggs by bird parents affect hatchling outcome. There is no evidence of which we are aware of parental manipulation of nest temperature to modify hatchling develop-ment. Rather, it is assumed that parents are trying to maintain eggs with a temperature range that produces the greatest likelihood of successful hatch-ing. But it is an interesting question whether variation in the mean value and consistency of egg temperature might result in differences in later life physiology in birds.

LIVE BIRTH

One strategy to protect eggs and offspring from potential predators and environmental conditions is to retain them within the female's body and produce live-born offspring: true pregnancy. Every vertebrate class—fish, amphibians, reptiles (but not birds), and mammals—has species that have evolved this strategy. Fossil evidence of live birth extends back to 380 million years ago, in large, apex predator lobe-finned fish called placoderms (Long et al., 2008). Live birth in the mammalian lineage, however, may only go back 160–200 million years. The fact that the monotremes lay eggs strongly sug-gests that lactation preceded live birth in the ancestors of mammals.

Some species that have evolved live birth do not appear to have evolved a mechanism for maternal transfer of resources to the developing egg/embryo, but rather rely on a large egg yolk sac. For example, coelacanths produce tennis-ball-sized eggs that are retained in the female's body until they have hatched. Coelacanths are not truly viviparous; animals with this form of egg retention with "live birth" are termed ovoviviparous. The parental care is providing a protected environment for the developing egg. In the case of coelacanths, who live in deep water, the eggs are protected from the extreme water pressure of those depths.

Other species have evolved mechanisms to transfer maternal resources to the retained egg/embryo. Some examples are discussed later in this chapter.

Providing Food

Many species, both vertebrates and invertebrates, have evolved mechanisms for providing food for their offspring. For example, there are species of spiders that provide food for their spiderlings. Often this food is simply captured prey (Ito and Shinkai, 1993), but in some species the mother re-gurgitates fluids on which the spiderlings feed (Schneider and Lubin, 1997). These fluids likely are a mix of digested prey and substances produced by the mother spider.

Parasitism is another mechanism for providing food for offspring. Parasitic wasps lay eggs on or in immobilized prey; the hatchlings feed on the often still-living host. The Jewel wasp has evolved an even more complicated and bizarre reproductive strategy. The Jewel wasp creates zombie cockroaches on which its offspring feed (Libersat and Gal, 2014). The mother wasp stings her cockroach prey twice; the first sting briefly immobilizes the cockroach for a few minutes. The second sting is directed to the head area. The wasp then leaves to find and prepare a nest site, a process that can take 30–40 minutes. When the cockroach recovers from the temporary paralysis from the first sting, it does not attempt to flee. Instead, it engages in grooming behavior for 30 minutes. When the wasp returns, it bites off the cockroach's antenna, grasps the cockroach by an antenna stump, and leads it to the nest site. The cockroach enters its tomb and then stays quiescent while the wasp lays her eggs and then blocks up the nest site.

Although these types of parental food provisioning are fascinating, demonstrating the power of evolution to create adaptive biology that reads like science fiction, this book is about milk and the mammalian maternal care strategy called lactation. Milk differs fundamentally from the previous examples of feeding offspring because milk is actually produced for that purpose by the mother's body. The rest of the examples of non-mammalian parental feeding in this chapter all involve offspring feeding on a substance actually produced by the parent, and not one from the external environment provided by the parent. In many of these cases, this parental behavior is actually a form of cannibalism.

Cannibalism

Cannibalism as a parental care strategy has evolved in many taxa and takes various forms. It does not always require the death of the parent, though in some cases that is what happens. For example, in an extreme case of maternal resource transfer found in at least six lineages of spider species, the female spider provides herself as a source of nutrition for her hatchlings in a form of suicidal maternal investment (Schneider, 1996; Kim et al., 2000). In some of these cases the matriphagy is facultative, not obligate; older, decrepit female spiders are the ones eaten by their last brood. In this case, the behavior can be considered to be parental care, since the offspring gain from eating their mother. But it is not really parental investment, as the female is sacrificing little if any future reproductive success, as she would not likely have ever produced another clutch of eggs. However, females of the species *Stegodyphus lineatus* apparently do engage in this extreme form of parental investment. If the female is removed from her spiderlings, and thus spared from being eaten, she will lay more fertile eggs in the future, and those eggs will develop into spiderlings (Schneider and Lubin, 1997).

A species of pseudo scorpion (*Paratemnoides nidificator*) provides food for nymphs in two ways. These pseudoscorpions are colony dwellers and cooperatively hunt, enabling them to capture prey up to four times their own size (Tizo-Pedroso and Del-Claro, 2007). The prey is shared with the nymphs. Under laboratory conditions, solitary female pseudoscorpions only produced a single clutch, while females in a colony with other females produced on average three clutches, suggesting a significant selective advantage to living in a colony (Tizo-Pedroso and Del-Claro, 2005). When prey is scarce, and a female has not captured prey for a few days, she offers herself as prey to her nymphs. She approaches her nymphs and spreads her claws wide, vibrating them. The nymphs gather and begin to feed on her body as she remains motionless (Tizo-Pedroso and Del-Claro, 2005). This sacrificial behavior is suggested to decrease the incidence of cannibalism among the nymphs, thus benefiting maternal reproductive fitness at the expense of the mother's own existence.

In the spider *Amaurobius ferox*, the mother is always devoured by her young shortly after their first molt. The cannibalism is preceded by increasingly frequent interactions between the female and her spiderlings, and it appears that the female plays a role in stimulating the behavior that leads to her death. However, if female *A. ferox* are removed from their spiderlings, some will manage to produce another clutch of viable eggs. In an experimental manipulation, female *A. ferox* mothers were either left with their young or removed (Kim et al., 2000). One-third of the females removed from their spiderlings went on to produce a successful second clutch of eggs. However, the number of spiderlings produced was fewer than for the first clutch. Spiderlings that ate their mothers more than doubled their body weights, went through their third molt sooner, and were larger than spiderlings that did not benefit from matriphagy at the time of dispersal. The survival rate was also higher; 94% of spiderlings that ate their mothers survived to the third molt compared to only 71% of spiders deprived of maternal nutrition. Eating their mother had clear benefits for spiderlings.

Did the mothers' reproductive fitness benefit from being eaten? In general the answer is yes. The average number of spiderlings surviving to dispersal was 82 for females that produced a single clutch and were eaten. The average number of successful young produced by females that were removed from their first clutch was 75 (62 from the first clutch and an average of 13 from the second). The better reproductive strategy for female *A. ferox* would appear to be to produce a single clutch of eggs and be eaten. However, the result may be a little more complex. Some uneaten females produced many fewer spiderlings (estimated 62 from the first clutch and none from a second clutch), while those females that managed a successful second clutch actually produced more (an estimated 101 surviving spiderlings, on average). The females that produced a viable second clutch were the larger females, suggesting that females in excellent condition might benefit, at least numerically, by leaving their first clutch and producing a second. Of course the spiderlings from their first clutch would suffer, and therefore their ultimate survival to reproduction may be compromised. In the end, producing a single clutch and performing the ultimate sacrifice would appear to be the evolutionary successful strategy for female *A. ferox*.

Not all cannibalism by offspring is performed on the parent. Another form of cannibalism as a reproductive strategy is siblicide. In 1948, a shark researcher was probing the uterus of a late-term pregnant sand tiger shark (*Carcharias taurus*) when he was bitten. The shark fetuses had teeth! Sand tiger shark embryos begin life with large yolk sacs; but these yolk sacs are not sufficient. Once they have used up their yolk sac they begin to feed on each other, a strategy termed embryophagy. In effect, a female sand tiger shark produces two types of offspring: successful ones and food for the successful ones. Eventually, a single offspring is left alive in each of the two uteri. By this somewhat grisly reproductive strategy a nine-foot long sand tiger shark can produce two three-foot long offspring.

Cannibalism occurs in spade foot toad tadpoles, but siblicide apparently is usually avoided. The tadpoles are polymorphic in nature, with some being detritus feeders while others become carnivorous. The two types develop morphologically different mouth parts. This polymorphic development is environmentally driven, based on the food the tadpoles consume shortly after hatching. In the laboratory, tadpoles fed detritus early on develop detritus-feeding mouth parts, while siblings fed on fairy shrimp develop the carnivore mouth parts (Pfennig, 1992).

Detritus feeders preferentially associate with kin, while carnivorous tadpoles tend to avoid kin (Pfennig et al., 1993). Carnivorous tadpoles eat other tadpoles but tend to spare their kin. They would nip at another tadpole and then consume it if it were not kin, but release it unharmed if it were kin (how they can tell is unknown). However, when they were food-deprived all bets were off and carnivorous spade foot tadpoles consumed kin and non-kin tadpoles at equal rates (Pfennig et al., 1993).

Siblicide also occurs in a number of bird species; however, cannibalism appears rare. Siblicide in birds appears more driven by competition for parental care, especially food being brought to the nest. Younger, smaller nestlings are killed by their nest mates, reducing the number of mouths the parents must feed. However, in some raptor species the carcasses of siblings killed in the nest are occasionally eaten (e.g., kestrels; Wiebe, 1996), and a Saker falcon chick was observed attacking and eating its still-live nest mate (Ellis et al., 1999).

Trophic Eggs

In many sharks there is a less grisly form of the embryophagy strategy, called oophagy. In other words, the developing embryos feed on eggs, not on other developing embryos. In some species these eggs may be fertilized, so it is still siblicide. However, in some species the mother continues to produce eggs that are not fertilized, or at least don't develop. In this way she produces a constant stream of nutrition for her embryos from her own reserves. The ovaries have switched from an explicitly reproductive function to become a provider of nutrients and probably other important molecules for offspring. Eggs produced as food for existing offspring, whether fertilized or not, are called trophic eggs. This reproductive strategy has evolved in many species, from spiders to amphibians. It is quite common in invertebrate species, especially the social species, such as ants.

Many spider species produce trophic eggs. The spider *A. ferox*, mentioned above as an instance of evolved matriphagy, produces a clutch of trophic eggs in the period just before her spiderlings undergo the first molt (Kim and Roland, 2000). Thus, she first produces trophic eggs for her offspring before becoming their meal herself. Lady bugs also produce trophic eggs. Approximately 10–25% of lady bug eggs never hatch; they are apparently infertile. Infertile eggs are randomly distributed in the clutch and can't be discerned from fertile eggs until the lack of development becomes obvious. The offspring that do hatch feed first on these eggs, before foraging for aphid prey. In an experimental manipulation, lady bug females were food deprived for 24 hours before they laid their egg clutch to simulate a low prey density environment. In agreement with prediction, the percentage of infertile eggs significantly increased, although total egg mass did not differ. It would appear that lady bug mothers are able to respond to a signal of potentially low density of prey items for their offspring by providing them with more initial food in the form of trophic eggs (Perry and Roitberg, 2005). Female lady bugs trade off the number of offspring produced with the survival rate of those offspring.

Some species produce trophic eggs that are distinct in morphology from reproductive eggs. In some cases the trophic eggs are unfertilized or somehow intrinsically infertile; in others, the eggs have specialized cellular development that renders them non-viable, even if fertilized. For example, the bug *Adomerus triguttulus* produces non-viable trophic eggs that are, on average, smaller than viable eggs. Importantly, these trophic eggs lack micropylar processes (necessary for sperm to penetrate the chorion of the egg) on their surface; viable eggs have 3–8 micropylar processes surrounding a pole (Kudo et al., 2006). Thus, trophic eggs of *A. triguttulus* are intrinsically non-viable, and likely represent an evolved adaptation for feeding offspring.

Frogs are known to provide both maternal and paternal care by carrying tadpoles on their backs; but only mothers are known to feed their tadpoles. In a number of species of frogs, mostly arboreal species, females return to the site where they laid eggs after the eggs have hatched into tadpoles and lay more eggs on which the tadpoles feed. The frog *Leptodactylus fallax*, native to the eastern Caribbean, deposits its eggs into a foam nest produced by fluid excreted by the female and whipped into a foam through vigorous paddling motions by the feet of the breeding male (Gibson et al., 2004). The nest is defended by both the male and the female. After the tadpoles hatch, the female continues to deposit eggs into the nest. The infertile eggs are about half the diameter of fertile eggs. She will deposit eggs into the nest every three days on average (range of 1–7 days). Tadpole guts contained identifiable eggs after the provisioning events (Gibson et al., 2004). The female reproductive system has become co-opted into a tadpole feeding mechanism.

Less-Extreme Cannibalism

There are examples of what is effectively cannibalism that doesn't result in death for the creature being eaten. Caecilians are tropical, legless amphibians; species range in size from as small as an earthworm to 1.5 meters long. They are burrowing creatures, with a strong skull and pointed snout well suited to pushing through mud and dirt. Their sight is limited to light-dark distinction. The most recent molecular studies point to a divergence of caecilians from the other amphibians more than 300 million years ago. They

Figure 1.1. Female *Boulengerula taitana* with young, which feed off her skin.
Photo: Alexander Kupfer.

are unusual among amphibians in that all species practice internal fertilization, and three-quarters of the species are viviparous (produce live-born
young).

The embryos of viviparous caecilians develop a special fetal dentition that
they employ to scrape the lining of the mother's oviduct; in essence feeding
on their mother before birth. In one species of egg-laying caecilian (*Boulengerula taitana*) the hatchlings are born with similar teeth, which they use to
feed on the mother's skin (figure 1.1). Brooding females produce a lipid- and
protein-rich, thick skin which her offspring peel off and consume (Kupfer
et al., 2006). Offspring appear to derive all of their nutrition from feeding on
their mother's skin until they reach the size at which they disperse from the
nest. *Boulengerula taitana* females produce the smallest brood size of any
caecilian (mean of 5 eggs; Malonza and Measey, 2005) with broods of between 2 and 9 young found by Kupfer et al. (2006). The young stay with the
mother for a few weeks; female body condition is negatively associated with
the length of time the young are in attendance, implying a substantial cost to
mothers (Kupfer et al., 2008). Indeed, in the laboratory, females lost 14% of
their body mass in one week of skin-feeding (Kupfer et al., 2006). However,
mothers do survive and can produce future offspring.

Secretions

In many viviparous species the eggs/embryos are nourished within the mother's body by uterine secretions, sometimes termed "uterine milk." These secretions may be absorbed through a yolk sac placenta or other specialized fetal tissue, or actually fed upon by the developing embryos. Uterine secretions are important to early fetal nutrition for placental mammals, including humans. Indeed, within the first 10 weeks of human gestation more nutrition and bioactive molecules are transferred to the placenta and fetus via uterine secretions than directly from the mother's blood (Jauniaux et al., 2003). Although uterine milk is far removed from mammary milk, there are possible functional and analogous similarities with ancient milk, which may have been a secretion deposited onto and absorbed by eggs (Oftedal 2002b).

SKIN SECRETIONS

Milk is a secretion from a specialized skin gland. Secretions from skin glands are common among vertebrates. In chapter 2 we explore the ancient origin of tetrapod (four-limbed vertebrates) skin glands, and the origins of the mammary gland in an ancient offshoot from those early amphibian-like ancestors. However, secretions from skin glands as food for offspring is rare, outside of mammals.

There are a few non-mammalian species that have evolved secretions on which their offspring feed. These adaptations are more analogous to milk. In discus fish, an Amazonian cichlid, both males and females produce mucosal secretions that the free-swimming fry feed upon (Hildeman, 1959; Buckley et al., 2010; figure 1.2). The secretions contain proteins and ions such as calcium, potassium, sodium, and chloride. A significant proportion of the protein fraction consisted of immunoglobulins (Buckley et al., 2010). Cortisol was not detectable in secretions from wild fish but was detected in aquarium-raised fish. Thus these mucosal secretions potentially are more than food.

Intriguingly, but not surprisingly, the production of these mucosal secretions appears to be linked to prolactin (Khong et al., 2009). Prolactin is an ancient molecule associated with a wide range of functions; but certainly prolactin is linked to many aspects of parental care and derives its name from its prolactational actions. The receptor for prolactin is found in the skin of discus fish, and its expression is upregulated in parents that are producing mucus for their fry (Khong et al., 2009).

Figure 1.2. Male and female discus fish with fry, which feed on skin mucus secretions from the parents. Photo: Mehgan Murphy, Smithsonian National Zoological Park, Washington DC.

We explore the evolution and functions of prolactin in more detail in chapter 4. For now we merely note that another major thesis of this book is that existing molecules are often co-opted to new function, using their existing properties and capacities to support novel evolved adaptations. Prolactin is linked with fluid balance, especially in fish, so it is not surprising that it should be involved in skin secretions. It is also an important molecule in parental behavior and, of course, lactation, as well as in our next topic: bird crop milk.

CROP MILK

Parental care is ubiquitous in birds. In many species it is bi-parental care, with both males and females providing body heat, protection, and food to their offspring. For most species, however, the food given to chicks is from the external environment. It is obtained, not produced. However, there are three types of birds that have evolved a parental feeding strategy that includes producing a secretion that is fed to the chicks: pigeons and doves, flamingos, and Emperor penguins. This secretion, often termed crop milk, is probably the closest analogy to mammalian milk found in non-mammals.

Among Emperor penguins, interestingly, it is the male that produces a form of crop milk from the esophageal lining, and then only a limited amount over a short duration immediately post hatching. Flamingos produce a secretion from glands along the whole upper intestinal tract which contains red and white blood cells in addition to fat, protein, and minerals. Flamingo crop secretions serve more than a nutritive function; transfer of maternal antibodies for West Nile virus to chicks has been documented in captivity (Baitchman et al., 2007).

Many species of Procellariiformes (albatrosses, petrels, and storm petrels [but not diving petrels]) produce stomach oil in the proventriculus which is fed to their chicks. Stomach oil is produced from dietary items; it does not appear to include proventricular secretions to any measurable extent (Place et al., 1989). Rather, it is a distillation of dietary lipids that are retained within the proventriculus; a fascinating anatomical adaptation, but not really crop milk.

The best-documented producers of crop milk are the pigeons and doves, which comprise the avian family *Columbidae*. Both male and female pigeons and doves produce a protein-rich secretion from glands in the crop that is fed to the chicks (Gillespie et al., 2012). Pigeon crop milk is mostly fat (60%) and protein (~35%), but does contain a small amount of carbohydrate in the form of oligosaccharides as well as minerals (Davies, 1939). Some of the protein in pigeon crop milk is actually immunoglobulins (e.g., IgA; Goudswaard et al., 1979) and peptide hormones such as transferrin (Frelinger, 1971) and pigeon milk growth factor (Shetty et al., 1992). Pigeon crop milk also contains microbes (Gillespie et al., 2012). Thus, pigeon crop milk appears to contain most if not all of the types of bioactive substances that mammal milk contains.

The non-nutritive substances in pigeon crop milk appear to be very important. If pigeon hatchlings (squabs) are fed an artificial diet nutritionally similar to crop milk instead of crop milk, they fail to thrive and even die (Guareschi, 1936). Newly hatched domestic chickens fed crop milk mixed into their regular feed grew larger both compared to control chicks fed only the regular diet and also compared to chicks fed a diet supplemented with an artificial crop milk that matched the fat and protein composition of crop milk (Gillespie et al., 2012). Body composition and relative growth also differed; breast muscle was enhanced in crop-milk-fed chicks. This strongly suggests that bioactive factors in crop milk that could regulate development and not nutrients underlie the differential growth.

The crop-milk-supplemented chicks differed from control chicks in their gut microbial community as well. Six microbial species found in crop milk were found only in the guts of chicks fed crop milk. Other microbes were found in all chicks, but in different proportions between crop-milk-fed chicks and control chicks (Gillespie et al., 2012). The oligosaccharides in crop milk may act as prebiotics, enhancing the growth of certain bacteria. This has been demonstrated to be the case for certain oligosaccharides found in human breast milk (Ward et al., 2006).

As in the example of the cichlid fish, above, prolactin is involved in this parental-care strategy. Prolactin provides a necessary signal to the tissue of the crop which then greatly enlarges and becomes a secretory organ (Goldsmith et al., 1981). This function of prolactin in these and other parental-care strategies including lactation are considered in more detail in chapter 4.

What about Dinosaurs?

Is it possible that some dinosaurs produced secretions or other bodily substances that they fed to their offspring? Birds are descended from a lineage of dinosaurs, and, as outlined above, some birds have evolved mechanisms to produce food for their chicks from their own bodies. The fossil evidence shows that many dinosaur species practiced parental care. There are assemblages of dinosaur fossils comprised of the full range of life stages: eggs, babies, juveniles, and adults. Communal nesting is common. Did any dinosaur species have offspring-feeding strategies analogous to crop milk, lactation, or any of the other strategies reviewed in this chapter?

Although there is no unambiguous fossil evidence to support parental provisioning by dinosaurs, the rapid growth rates of some dinosaur species suggests that parental provisioning of food likely occurred. It would be surprising if some form of parental care that included food provisioning didn't evolve in some dinosaur lineages. It should be the default assumption for the avian lineage of dinosaurs, given the ubiquitous nature of this kind of parental care in birds. We don't know if that food was produced by the parents or merely regurgitated. At this time all of this is fascinating speculation, and we direct the reader to a short paper that outlines what case can be made for any lactation-like adaptations in dinosaurs (Else, 2013).

Not Just Nutrition

The above examples make it clear that strategies to feed offspring from parental secretions or tissue are widespread among animals. Many of these adaptations are quite analogous to milk, though none is homologous. Milk is unique, and only mammals can be defined as a taxonomic unit by a single, common strategy: producing milk. That is truly one of the unique aspects of milk and lactation; all living members of an evolutionary lineage that stretches back hundreds of millions of years continue to employ this ancient adaptation.

One important aspect of milk that we emphasize in this book is its function as a signaling/regulatory mechanism that allows mothers to direct the development of aspects of their offspring's physiology and behavior. Many of the non-mammalian adaptations also have the possibility of serving these functions, though few have been examined in enough detail to ascertain how likely that would be. Trophic eggs certainly have the potential to contain tremendous amounts of biochemical information in addition to nutrients. An interesting area of potential research is to what extent trophic eggs vary in biochemical composition in response to the external or internal environment of the mother or to the age of the offspring. Some, but not all, mammals produce milk that changes in composition as the offspring age. This is especially true of marsupials. Do sharks that produce trophic eggs for their pups always produce the same eggs, or does the composition change over time?

The mucus secretions of discus fish contain immunoglobulins that assist the offspring in warding off disease; do they contain other signaling molecules in functional concentrations? Do these secretions change over time? Or change in response to some stimuli that affects the parents? Or are these secretions invariant? Crop milk contains hormones, immunoglobulins, and microbes in addition to nutrients; the full complement of signaling molecules found in mammalian milk. The evidence supports crop milk playing an important role in regulating growth and development beyond that from nutrition.

Cells versus Secretions

One difference between milk and most of the adaptations of other species reviewed here is that most of the foods fed to non-mammalian offspring contain cells. Trophic eggs, hypertrophied skin, even crop milk is primarily composed of cells. Milk certainly contains cells, both of maternal origin and microbial, but milk itself is not made up of cells. It is a secreted fluid that has evolved to be food.

Cells, by definition, contain the necessary constituents of life. After all, they are alive. Food for your offspring composed of cells from your body would likely have all the nutrients, signaling molecules, and other necessary bioactive molecules to support life, growth, and development. It is more unusual to have a secretion that contains all of these things. There are foods in nature that do not contain cells; for example, nectar from flowers. But these foods are not complete. Animals cannot thrive on them alone. Milk is perhaps unique in this attribute (the mucus secretions of cichlid fish are at least analogous, though it is not known if they are sufficient food for the fry).

There are potential advantages and disadvantages to a secretion as food rather than a cellular substance as food. The advantage of cell-based food is that it will generally be complete. A disadvantage is that the range of variation will be constrained by the necessity of the cell being alive. There will be limits on the concentrations of bioactive molecules within the cell. A secreted substance, such as milk, presumably could go outside those limits. The composition of milk is not like that of a cell. The variation in milk composition far exceeds the variation in cell composition between mammals.

Chapter 2

Origins

Lactation and milk were one of the earliest challenges to Darwin's theory of evolution by natural selection. In his book, *On the Genesis of Species,* St. George Jackson Mivart presented a scholarly critique of Darwinian evolution, arguing that many features of animals were too complex and their function so reliant on that complexity for them to have arisen from gradual changes to simpler structures that lacked much functionality. Milk was one of his examples. He asked how milk could have evolved due to benefits "from a scarcely nutritious fluid from an accidently hypertrophied cutaneous gland" (Mivart, 1871; p. 53). It is quite a challenge to envision how something as complex as lactation, producing a biochemically complex substance such as milk, could arise by gradual, step-by-step evolutionary processes.

One of the flaws in Mivart's reasoning is that he presumed that the current function of milk was the same as the original function in the distant past. Function evolves. Evolution acts on existing adaptations to create new functions. Although lactation is a uniquely mammalian trait, it has an ancient origin, and began in creatures that were not mammals. Based on analyses of the many novel constituents in milk, ancestral lactation appears to have originated in the Permian, perhaps as long as 300 million years ago (Oftedal 2002a; 2002b; 2013). Mammals don't appear in the fossil record until more than 100 million years later, in the late Triassic. The lineages leading to modern mammals (monotreme, marsupial, and eutherian/placental mammals) came into existence around 160 to 200 million years ago.

Because of the similarities among all mammal milks, whether from a monotreme, a marsupial, or a placental mammal, milk as a complex biochemical fluid likely has existed for more than 200 million years; but that still leaves 100 million years of evolution from its original form. There is no reason to presume that the function of the original skin secretions produced by the ancestral creatures from which mammals and modern lactation derive was to be food for their offspring. That is a mammalian adaptation. The secretions from a proto-mammary gland 300 million years ago could have served other adaptive functions, most likely during reproduction, as it is difficult to envision an origin for lactation that was outside of reproduction. However, although the nutritive function of milk is of primary importance for modern mammals, we know that milk has many other functions including immunological, anti- and pro-microbial, as well as regulatory. It isn't clear what the original primary function of the ancestral secretions from which milk derives might have been. Over time milk evolved to become a food for offspring; but it didn't necessarily begin as food, and even today it is more than a food.

To imagine the original adaptive function of secreting a fluid from a hypertrophied cutaneous gland, we have to imagine the possible reproductive lives of animals that no one would call a mammal. Dr. Olav Oftedal has done just that, providing a compelling narrative regarding the possible original functions of proto-lactation in early synapsids (the ancestors of mammals) 250–300 millions of years ago (Oftedal, 2002a; 2002b; 2012). We outline his thesis and other theories regarding the origin of lactation later on in the chapter, but we encourage the reader to explore his publications for more detail. First, to set the stage, we provide a general overview of the theories and evidence regarding how mammals came into being.

Life on Land

In the late Devonian, the first vertebrates began living at least part of their life on land. These first land vertebrates were tetrapods (four-limbed vertebrates) from about 365 million years ago and were descendants of the lobe-finned fishes (Carroll, 2009). They were the ancestors of all living terrestrial vertebrates, including amphibians, reptiles (which includes birds), and mammals. The early land tetrapods were still linked to water; many were fish eaters, and all likely deposited their eggs in water. Reproduction was similar to that of modern amphibians, where an egg develops into a larval stage first (e.g., a tadpole), before developing into the final adult form (Schoch, 2009). The word amphibian derives from the Greek meaning double life, and refers to this larval-adult stage dichotomy.

These amphibian-like tetrapods thrived through much of the Carboniferous era. Land plants were fully established and trees (giant club mosses, tree ferns, and giant horsetails) emerged to dominate the landscape. The environment was swampy rainforest, with plenty of water, high atmospheric oxygen content (about 30%), and atmospheric carbon dioxide levels many

Figure 2.1. Artist rendering of the giant millipede in the genus *Arthropleura* from approximately 300 million years ago. Artist: I. Sailko.

times what they are today, resulting in a warm, humid environment, eminently suitable for the amphibian lifestyle. The high oxygen content allowed giant insects to evolve; dragonflies the size of seagulls, meter-long cockroaches, and the largest known land invertebrates, millipedes in the genus *Arthropleura*, which could grow to be more than 2.5 meters in length and fed on the detritus on the forest floor (figure 2.1). The swampy rainforests extended across much of the land surface in a largely continuous swath from what is now the Midwest of the United States to Kazakhstan and even into China. These Carboniferous rainforests are also called the coal forests, as they formed large peat bogs that became the substantial coal deposits found in all these regions of the world. These coal deposits and the fossils within provide a fascinating window into this ancient world (box 2.1).

Many Carboniferous amphibians began to spend more time on land, hunting and feeding, but they still required water for reproduction; their eggs would desiccate if laid on dry land and their larval stages generally needed to live in water. During the late Carboniferous, around 320 million years ago, a lineage of tetrapods evolved an important reproductive adaptation that allowed fully terrestrial reproduction—the amniotic egg (box 2.2). The amniotic egg contains three extraembryonic membranes that served to allow the eggs to become larger (i.e., contain more maternal

BOX 2.1

The coal forests

Much of the coal found in the world was formed from ancient peat mires or bogs that were buried under extensive amounts of soil due to flooding and other natural processes. The high pressure and heat the buried peat was subjected to results in compaction and chemical changes called coalification. The swampy rainforests and shallow seas of the Carboniferous era were ideal for producing extensive and deep peat bogs that could be inundated and then covered in silt, producing the worldwide coal seams from this era. Plant fossils from coal deposits have long been an important window into the ancient past, informing the ideas of the eminent geologist Charles Lyell, the founder of geographical botany Joseph Hooker, and Charles Darwin.

The ceilings of many coal mines provide a fascinating glimpse into the ancient past. Deep coal mines in the United States often employ a "room and pillar" extraction method in which only part of the coal seam is removed, with pillars left to hold up the roof. The ceilings of these mines often contain extensive assemblages of plant fossils, reflecting the forest floor of the Carboniferous era (Falcon-Lang et al., 2009). Entire landscapes are in effect preserved, allowing paleobotanists to investigate the variation in plant composition across a large swathe of a 300-million-year-old swampy rainforest, providing information on plant community structure. These ancient peat mires were huge, covering hundreds of thousands of square kilometers in some cases (Falcon-Lang et al., 2009).

Evidence from coal plant fossils casts doubt on the notion that the Carboniferous tropics were continuously covered in rain/swamp forest. The polar ice cap on Gondwana expanded and shrank in a cyclical fashion related to changes in solar radiation due to what are known as Milankovitch orbital cycles. During the warm periods, sea level was higher and the Carboniferous tropics were wet, warm, and extensively covered in swampy rainforest—ideal conditions for the production of the extensive peat bogs that eventually became coal. But during the glaciation periods, the tropics were cooler and dryer, the rain forests shrank, and extensive areas dominated by conifers and other arid-adapted plants existed. The transition from the Carboniferous to the Permian was not a simple change of decreasing rainforest and increasing arid habitat, but rather a cyclical pattern of rainforest expansion and decline and an associated decline and expansion of arid habitat, with the extent and duration of rainforest dominance decreasing over time. By the end of the Carboniferous, the rainforests had become restricted to relatively small refuges, with one enclave in China surviving into the late Permian.

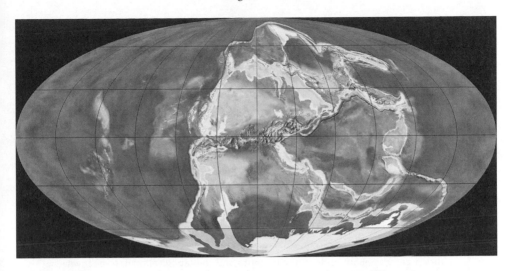

Figure 2.2. The supercontinent Pangea approximately 280 million years ago.
© Ron Blakey, Colorado Plateau Geosystems, Inc.

resources), to resist desiccation, and to store metabolic waste. A larval stage of development outside the egg was no longer required due to the extra resources deposited into the egg, and the eggs did not need to be deposited in water.

This was a time when the continents were colliding. Laurasia in the northern hemisphere was moving south, colliding with Gondwana moving north, forming the supercontinent Pangea (figure 2.2). The formation of Pangea corresponded with a dramatic change in climate, to a much drier, even, desert-like landscape for most of Pangea in the Permian era (Tabor, 2013). About 305 million years ago, the rainforests covering the supercontinent Pangea began to die off rapidly, in what is called the Carboniferous rainforest collapse.

The change from a landscape dominated by large tracts of swampy, wetland forests to isolated rainforest tracts separated by extensive arid landscapes had a profound effect on the terrestrial vertebrate communities. Local biodiversity generally decreased, as many species went locally extinct. However, the number of endemic species (species found only in a limited locale) increased due to the isolation of the rainforest fragments, leading to an overall increase in biodiversity. At the peak of the Carboniferous rainforests, the same species could be found over very wide ranges. When the rainforests were broken up into islands of forest in a sea of

BOX 2.2

The amniotic egg

Amniotes are vertebrates that produce amniotic eggs. What distinguishes the amniotic egg from fish and amphibian eggs is a unique set of membranes: the amnion, the chorion, and the allantois. These membranes are multicellular, vascularized tissues that grow out of the embryo and thus are referred to as extraembryonic membranes. The amnion surrounds the embryo and creates a fluid-filled cavity in which the embryo develops. In humans and other placental mammals, this becomes the amniotic sac and is filled with amniotic fluid. The chorion forms a protective membrane around the egg. The allantois is a special membrane that performs gas exchange and stores metabolic wastes from the embryo; eventually it forms the bladder. In placental mammals these two extraembryonic tissues, the chorion and the allantois, form the placenta and the umbilical cord, respectively. The antecedents of the modern mammalian placenta date back to adaptive changes in egg structure that occurred during the Carboniferous era, more than 300 million years ago.

What were the adaptive advantages provided by amniotic eggs? The extraembryonic membranes of the amniotic egg protect the embryo from desiccation. The amniotic egg also allowed for greater transfer of maternal resources; eggs could become larger now that they had the protective chorionic membrane and the allantois to handle metabolic waste. More maternal resources could be placed into the egg, and the embryo had greater protection from the external environment. The suggested water-loss prevention function of the amniotic egg matches with other adaptations in amniote biology that also appear to prevent desiccation. For example, the skin of amniotes is relatively impervious, contrasting with the skin of amphibians, through which water and gases easily flow. Amniotes possess a high density of renal tubules and a water-resorbing large intestine. Much of amniote biology can be viewed as adaptations to support fluid balance, which enables them to live on land, away from water.

desert, each forest island began to develop its own species community. The result was more total species, but most species had restricted geographic ranges. There were now more laboratories in which evolution could work, though each one could not support the breadth of species found in the huge, continuous forests of the past. In addition to the challenges presented by the growing expanses of arid environments, there was also opportunity. Species that could survive away from open water would have advantages. It may have been during this time that a lineage of synapsid amniotes evolved a cutaneous secretory gland that eventually became the mammary gland.

Fluid Balance

Regulating fluid balance is a fundamental challenge for terrestrial animals. Life originated in water, and water is necessary for life. In order to invade land, animals had to solve the problem of water loss. The first tetrapods that invaded land appear to have been scaled, similar to many lobe-finned fish. But scales are not an effective barrier to water loss (Licht and Bennet, 1972; Roberts and Lillywhite, 1980; Lillywhite, 2006). Other adaptations arose to reduce water loss through the skin (box 2.3). An important adaptation linked at least in part to regulation of fluid balance and vital to the evolution of lactation was glandular skin.

The increasing aridity after the collapse of the Carboniferous coal forest favored terrestrial species with enhanced adaptations against water loss. Many of the amphibian-like species went extinct during this period, likely due to the loss of the wet habitats they required. But a particular offshoot of the early terrestrial tetrapods, the amniotes, fared better, most likely due to their adaptations for water balance and reduction of water loss. There are two surviving ancient lineages of amniotes: the synapsids and the sauropsids. Mammals are the only living descendants of the synapsids. The sauropsids gave rise to lizards, snakes, crocodiles, turtles, tuatara, and dinosaurs and their descendants, the birds.

With the rise of the amniotes, vertebrates that could live their entire life cycle outside of water existed. The amniotic egg, with its extraembryonic membranes, protects the egg from desiccation and thus enables it to be laid on dry land. Amniotes were better adapted than their amphibian-like tetrapod ancestors to arid conditions. Indeed, many of the adaptations in both surviving lineages of amniotes appear related to alleviating water loss.

It is therefore interesting to consider that lactation increases water loss from the mother but provides an important source of water for her offspring. Fluid balance may have been both an important constraint as well as an adaptive function for early lactation.

Synapsids

Mammals are the only living descendants of the synapsid lineage of amniotes, which originated more than 300 million years ago in the Carboniferous

BOX 2.3

Adaptations to reduce water loss through skin

The invasion of land by vertebrates is associated with a number of changes in integument or skin. Most of these changes are thought to be in response to the different selective pressures land dwelling imposed. An aquatic existence reduces the mechanical stress due to gravity through buoyancy; the water holds up the animal against gravity. A primary function of skin is protection, such as from mechanical damage (e.g., predator teeth), pathogen attack, sunlight, and other environmental challenges. On land the animal must support its entire weight. It is no accident that the largest animals ever to live have been aquatic (e.g., the blue whale). Exposure to sunlight is greater, as there is no water layer to absorb UV light. Living on land produced novel and enhanced challenges to skin.

Regulating the flow of water in and out of the animal is a vital function of skin for all vertebrates, whether aquatic or terrestrial. The invasion of land is linked to a number of adaptations and instances of evolutionary bursts regarding structures and molecules important to skin that affect water transport. In the aquatic fish and amphibians, organelles related to barrier function are composed of mucus and glycoproteins. In amniotes, they are composed of glycolipids and lipids (Lillywhite, 2006). The resistance to water passage by skin (R_s), measured in units of seconds per centimeter, generally ranges from 1–10 s/cm in amphibians and 30–500 s/cm in amniotes. Vertebrates living in xeric environments can have much greater R_s, with even amphibians achieving resistance of 200–400 s/cm, and in reptiles greater than 1,000 s/cm.

An important evolutionary event was the duplication/diversification of keratin protein genes. There are multiple keratin genes in all tetrapods, and the timing of the keratin gene expansion suggests that the gene multiplication occurred coincident with land invasion. All amniotes have multiple copies of the type I and type II α-keratin genes. The β-keratins are only found in the sauropsid lineage; all the reptiles including birds (Vandebergh and Bossuyt, 2012). The β-keratins are not related to the α-keratins, and appear to derive from an evolutionary event in the sauropsid lineage sometime soon after it diverged from the synapsid lineage leading to mammals. Indeed, some authors suggest they should be renamed, as they are not true keratin proteins (Strasser et al., 2015). The keratin proteins perform multiple functions, but many are important in providing structural integrity and reducing the permeability of the skin.

era and rose to dominance during the Permian period, between 260 and 286 million years ago. Synapsids are defined by the existence of a single temporal fenestra (hole) behind each eye orbit. These fenestrae served as attachment points for the jaw muscles and provided adaptive advantages in chewing. Synapsids also had differentiated teeth; that is, they had more than one kind of tooth in their mouth. *Dimetrodon,* an apex-predator synapsid from the middle Permian (figure 2.3), had two kinds of teeth—shearing teeth and sharp canine teeth. Most, but not all, modern mammals have four tooth types: incisors, canines, premolars, and molars. Both of these adaptations, jaw muscle attachments and differentiated teeth, are hypothesized to have provided significant adaptive advantages by reducing food into small particles before swallowing, increasing the efficiency of digestion and the rate at which nutrients could be extracted from food. Effective mastication is an important adaptive feature allowing many mammalian species to maintain their high metabolic rate relative to the many taxa from the sauropsid branch of amniotes. Of course there are counter examples; birds are toothless descendants of sauropsids that have metabolic rates equivalent to most mammals, and there are toothless mammals (though these species, mostly anteaters, do tend to have metabolic rates lower than the mammalian average). Still, it is reasonable to hypothesize that the early synapsid mastication adaptations may have allowed a higher rate of energy turnover, including a high basal metabolic rate which allowed them to be endothermic (warm-blooded), and those characteristics translated into a competitive advantage.

During the Permian, synapsids occupied a large number of niches. There were large and small synapsids—carnivores and herbivores. Throughout the middle Permian, the most common synapsids were the pelycosaurs, the dominant vertebrates of the Permian ecosystem. This clade of synapsids gave rise to the therapsids in the late Permian. The therapsids replaced the pelycosaurs as the dominant land vertebrates until the Permian-Triassic extinction event. During the late Permian, a group of therapsids known as cynodonts arose and were eminently successful. Some cynodonts survived the Permian-Triassic mass extinction. Based on aspects of jaw and tooth morphology, mammals were determined to be descended from a lineage of cynodonts.

Synapsids produced what are called parchment-shelled eggs, which were potentially subject to greater moisture loss than the calcified eggs that evolved in some members of the sauropsid branch of amniotes (Oftedal,

Figure 2.3. Dimetrodon, a top-predator synapsid from approximately 280 million years ago. Displayed at the Smithsonian Institution National Museum of Natural History, Washington DC.

2002b; 2012). If early synapsids laid their eggs in a moist environment, then the water loss through the parchment-shell may have been minimal, and, indeed, water absorbed from the environment may have been adaptive and even necessary for the fetus within the egg to develop. The increased aridity of the environments from the late Carboniferous and through the Permian would have reduced the environments in which burying eggs in moist ground would have been possible. Many synapsids would have been forced to lay their eggs in potentially desiccating environments. Adaptations to reduce water loss from eggs would have been favored.

There are many possible solutions to reducing water loss from eggs. Some sauropsids developed a calcified eggshell, which serves to reduce water loss in part by reducing the pore size of the egg (Oftedal, 2002a; 2013). Birds retain a hard, rigid, highly calcified eggshell, along with many other reptiles such as the crocodilians and many lizards. Other reptiles have leathery, flexible eggs, which are usually buried in moist ground and can absorb water. For example, the eggs of the king cobra can increase in weight by about two-thirds from absorbing moisture from the soil. But even these leathery eggs have a calcareous layer, though, of course much reduced in the amount of calcium relative to a bird egg. Even the eggs of reptiles that produce live-born young by retaining the eggs within the mother have the structures for a calcareous layer.

It would appear that a calcified shell was an adaptation evolved only in the sauropsid lineage. There is no evidence for any synapsid to have ever produced a calcified egg. Certainly the eggs of the monotreme mammals are not calcified. In fact, there is very little evidence for synapsid eggs in the fossil record; and none from the early years of their evolution. A calcified egg is much more likely to fossilize; witness the large numbers of fossil dinosaur eggs that have been found. The paucity of fossilized synapsid eggs is evidence that synapsids always produced parchment-shell eggs that would not be water resistant. The eggs could absorb water, if they were laid in a moist environment, but would lose water in a drier environment. The implication is that synapsids should have evolved behaviors (e.g., burying their eggs in moist soil) and/or other capabilities to ensure that their eggs developed under moist (or at least not dry) conditions. This leads to the milk-for-eggs hypothesis (Oftedal, 2002b). Or, more correctly, the original adaptive advantage of lactation was in providing water to eggs to maintain fluid balance when eggs were laid in a relatively dry environment.

Live Birth

There is another obvious possible solution to the problem of protecting an egg from water loss: retaining it within the mother's body. Live birth through egg retention is an ancient adaptation, going back at least 380 million years ago in large apex-predator lobe-finned fish (Long et al., 2008). Since tetrapods are descendants of a branch of lobe-finned fish it is reasonable to presume that the necessary genetics for egg retention and live birth existed in early tetrapods, though there is no evidence of live birth in any of these ancient animals. However, live birth has arisen independently in almost all surviving lineages of tetrapods. There are live-bearing amphibians (the caecilians), live-bearing reptiles among both snakes and lizards, and of course almost all mammals.

Is it possible that there were early synapsids that produced live young? Considering how many times among diverse lineages live birth has evolved it would not be at all surprising that a branch of early synapsids might have evolved this solution to protecting their eggs from water loss (and many other dangers). Just as apex-predator lobe-finned fish retained their eggs and produced live hatchlings, perhaps apex predator pelycosaurs like *Dimetrodon* did as well.

At the very least the reproductive strategy of extended embryo retention may have evolved. One possible strategy to reduce water loss from eggs is to reduce the amount of time the egg is exposed to the environment. The retention of eggs, or extended embryo retention, would have reduced the amount of time that the eggs were exposed to the more arid conditions after the Carboniferous. For example, a platypus egg takes about four weeks to travel through the oviduct, increasing in size as it travels. The eggs hatch about 10 days after they are laid. Early development occurs within the mother's body, with the egg spending almost three times as much time inside the female. In contrast, the chicken egg is laid after only about 24 hours of development inside the mother, and then she broods it for about 3 weeks.

In theory, embryo retention or live birth makes sense for early synapsids, but there is no fossil evidence it ever evolved. No embryonic pelycosaurs or other early synapsids have been found within adult fossils. Of course there is no fossil evidence of *Dimetrodon* or other early synapsid eggs either; but that likely results from the low probability that a parchment-shell egg would fossilize. The weight of the evidence favors the hypothesis that early synapsids laid parchment-shell eggs, and that live birth was an adaptation of therian mammals (the marsupial and placental mammals), occurring 100–150 million years or more after the origin of lactation.

Origin of Mammals

The earliest lactating animals were most likely a branch of the synapsid lineage of amniotes. They were not mammals; they were the ancestors of mammals. They produced a fluid that was the ancestor of mammalian milk. We suggest that this ability to produce a secretion important to both eggs and then hatchlings was a fundamental reason these ancestors of ours survived the Permian-Triassic mass extinction.

Approximately 250 million years ago, between the Permian and Triassic periods, a worldwide mass extinction event occurred. It is estimated that up to 70% of the existing land vertebrate species became extinct, including most of the synapsids and sauropsids. Although cynodonts survived into the Triassic, they were no longer as plentiful as they were in the late Permian. In addition, these descendant species were smaller on average than their ancestors and appeared to have become mostly nocturnal. The descendants of large, apex predators had become small, nocturnal insectivores.

This shift to a nocturnal lifestyle may have been a key event in the evolutionary history of mammals. Some researchers suggest that this specialization in nocturnal living selected for a higher metabolic rate and the ability to regulate body temperature by internal rather than external means. Because of their nocturnal habit, most of the synapsids of the dinosaur era could not raise their body temperature by basking in the sun. Instead, metabolic mechanisms to regulate body temperature evolved. Insulation against thermal losses would also be important. This could be accomplished by behavioral means, for example, by burrowing, nest-building, or huddling in a group. Some cynodonts in the early Triassic were burrowing animals. In one fossil bed, multiple individuals were found to have died, possibly from a flash flood, in the same burrow system, implying that these were social animals. Although not proof of the existence of parental care, social living is associated with parental care in many taxa. Hair or fur would also have a thermal insulation function; as of 164 million years ago, animals with fur existed (Ji et al., 2006). Hair is also associated with mammary glands and nipples (see below).

Milk for Eggs

The current evidence supports the hypothesis that lactation most likely evolved in an early branch of the synapsid lineage of amniotes. These early lactators were fully adapted to living their complete life cycle on land, and they were egg-laying animals, producing parchment-shell eggs. Fluid balance was a critical challenge at all life stages, including for embryos within the egg. Perhaps an original function of lactation was to combat the challenges to fluid balance of eggs from being laid on land, and thus in a dry environment. The sauropsid amniotes adapted to this challenge by evolving a calcified shell; according to the milk-for-eggs hypothesis, at least one branch of the synapsids apparently leaked fluid on their eggs.

The suggestion that the original function of lactation and milk (in whatever form it took) was primarily or even solely for eggs rather than for neonates has been convincingly developed by Dr. Olav Oftedal. Other scientists have provided intriguing suggestions along these lines, but Dr. Oftedal's work is the most rigorous and well developed. For the details of his fascinating theories of the origins of milk, lactation, and the mammary gland we recommend his recent publications (Oftedal, 2002a; 2002b; 2012; 2013). Here we briefly review his ideas about the earliest functions of milk.

An original function of proto milk (it might not have looked much like milk) was to bathe eggs in a fluid to reduce water loss, or even to provide additional water to the eggs (Oftedal, 2002b). After the Permian-Triassic extinction event, surviving synapsid species became progressively smaller in body mass. Smaller mothers imply smaller eggs. A small egg has a greater surface-area-to-volume ratio increasing the potential for water loss; but also increasing the potential for gain of water (and other substances) from fluid secreted onto the shell. Small eggs (or rather, the developing embryos inside) may have benefited more from maternal secretions than larger eggs, which may have been more buffered from water loss by the reduced surface area to volume and would have contained greater amounts of nutrients relative to need.

Lactation, Milk, and Survival

The Permian-Triassic extinction event decimated the synapsids. But some lineages survived, and, most importantly, the lineage that led to mammals survived. Was the adaptation of proto lactation, at whatever stage it had developed, a vital factor in the survival of our synapsid ancestors? Did the challenging circumstances of the Permian-Triassic extinction event serve as selective pressure to enhance proto lactation?

Figure 2.4. A reconstruction of *Hadrocodium wui*, an early Jurassic fossil. Reconstruction artwork: Mark A. Klingler/Carnegie Museum of Natural History for Luo et al., 2001. Reprinted with permission from AAAS.

The small size of many of the surviving synapsids (figure 2.4) suggests that lactation was a crucial aspect of their reproductive strategy. They would not have been able to produce a large enough egg for direct development to occur (Oftedal, 2013). Hatchlings likely would have been born altricial and required parental care for thermal regulation and provisioning of food. Extending the transfer of maternal resources via these secretions to hatchlings likely occurred early on. Oftedal (2012) argues that the small eggs could not have contained enough yolk to produce precocial hatchlings capable of thermoregulation, and certainly not produced hatchlings that could feed on food items similar to those in the adult diet. This implies a long evolutionary history of maternal care in the mammal lineage. The incorporation of nutrients into mammary secretions eventually allowed milk to supplant egg yolk as the primary maternal resource in the rearing of young.

Preadaptations for Lactation

In order for lactation to have evolved, certain characteristics needed to already exist in the early synapsids. They needed to have glandular skin, since the mammary is a modified skin gland. Another necessary preadaptation is maternal care, at least of eggs. There were undoubtedly many other pre-existing adaptations that created favorable conditions for lactation to have evolved. Even though lactation was a novel adaptation, its building blocks of anatomy, physiology, hormones, and behavior already existed. Evolution often works by co-opting existing features into new functions.

Glandular Skin

The early amphibian-like tetrapods had glandular skin, as do modern amphibians. Modern amphibians have mucous glands, which secrete mucous, and granular glands, which secrete bioactive molecules, including antimicrobial peptides. Thus, two important functions of amphibian skin gland secretions are water balance (mucous reduces water loss through the skin) and defense against microbial pathogens. Glandular skin is an early tetrapod adaptation, predating the amniotes and the synapsid-sauropsid split. There was a change in glandular secretions in the amniote lineage. The stem tetrapods secreted mainly mucus and glycoproteins; amniotes secrete glycolipids and lipids (Lillywhite, 2006). The water barrier function of skin in amniotes is in part achieved via Golgi-derived lipid-enriched secretory organelles (lamellar bodies) made up of glycosphingolipids, free sterols, and phospholipids. The contents of these organelles are secreted into the extracellular space where they form lipid bilayers within corneocytes within the stratum corneum, the outermost layer of epidermis (Lillywhite, 2006). This provides an enhanced water barrier, reducing water loss through the skin.

The ability to secrete lipid via skin glands could be considered a preadaptation for lactation that includes high-fat milk. However, the process of lipid secretion by Golgi-apparatus in mammary epithelial cells is unique and does not resemble the lipid secretion of other skin glands (Oftedal, 2013).

In the sauropsid lineage of amniotes (reptiles, including birds) the glandular component of skin has become reduced, probably due to the evolution of a novel protein (β-keratin) that is an essential component of water-resistant scales and feathers (Oftedal, 2013). In contrast, synapsids retained highly glandular skin, with many types of glands. Since the mammary gland is a modified cutaneous gland, glandular skin was a necessary preadaptation for lactation.

Parental Care

The existence of maternal care would seem to be a necessary condition for the evolution of lactation. If early female synapsids laid their eggs and then left, never to interact with eggs or hatchlings again, lactation would never have evolved. The evidence for parental care extends well back in time. Fossil evidence from 260 million years ago suggests parental care of some sort in a synapsid. An assemblage of five fossilized members of a pelycosaur species strongly suggests an adult with four juveniles, at the least indicating that adults and young associated (Botha-Brink and Modesto, 2007). Parental care is widespread and ancient. For lactation to have evolved maternal care would have been a necessity.

Origin of Mammary Glands

Mammary glands appear to have evolved from apocrine glands (Oftedal, 2002a; 2012). Mammary glands are associated with hair, and with hair follicles. Generally in mammals an apocrine gland is associated with a hair follicle and a sebaceous gland, forming an apo-pilo-sebaceous unit. In monotremes, the mammary patch is comprised of analogous (or probably homologous) associations of mammary glands with hair follicles and sebaceous glands in mammo-pilo-sebaceous units. In some marsupial mammary glands (e.g., those of opossums and kangaroos), hair follicles and sebaceous glands are also linked in the development of the nipples. The hair follicles of the mammo-pilo-sabaceous units actually penetrate the nipple epithelium during development, but the hairs are then shed. Each shed hair follicle leaves behind a duct (called a galactophore) which connects the mammary gland with the surface of the nipple. Monotremes have no nipples. The milk is wicked off the mother's skin by the hairs of the mammary patch, and the offspring lick the milk off the hairs. Hair and lactation, two defining mammalian characteristics, appear to be intrinsically linked.

Recent evidence on the regulation of cell development that links hair follicles to mammary cells is a fascinating glimpse into what may have occurred a few hundred million years ago. The direct association between mammary glands and hair follicles has been lost in the placental mammals (Oftedal, 2012). However, research in transgenic mice has shown that changing signaling pathways can change nipple epithelia to hair-bearing epithelia (Mayer et al., 2008) and can change hair follicles into glands with mammary gland–like activity (Gritli-Linde et al., 2007).

The Importance of Nipples

The evolution of the nipple was a major advance for the complexity of lactation. It allowed neonates to attach to their mother. The complex lactation of the marsupials, where a neonate is born in a fetal-like state (compared to a placental mammal neonate), crawls up to its mother's pouch, and then feeds on milk alone for a substantial period of growth and development doesn't work without a nipple on which the neonate can attach. It can be argued that the evolution of the nipple allowed the evolution of live birth in mammals. Certainly the only mammals that lay eggs lack nipples.

The ontogeny of the mammary gland and nipples are an important and fascinating aspect of the origin of lactation. It is also beyond what we can manage to discuss in this book. We direct the reader to Oftedal's many, comprehensive treatises (e.g. Oftedal 2002a; 2002b; 2012; 2013).

Mivart Revisited

We now return to Mivart's challenge to Darwin and evolution regarding that ancient "scarcely nutritious fluid from an accidently hypertrophied cutaneous gland." The evidence strongly supports that the first milk-like substance was produced by an early amniote on the synapsid branch, soon after the split between synapsids and sauropsids. These ancient ancestors of ours were fully adapted to life on land, laid parchment-shell eggs on land, and the female, at least, tended the eggs. Parental care may have extended to hatchlings as well, but the function of the proto-milk probably was primarily directed at the eggs. The female likely spent a significant amount of time in physical contact with her eggs, providing protection, body heat, and moist secretions from specialized glands on her ventrum (proto-mammary glands). These secretions provided water and minerals (e.g., calcium and potassium), as well as anti-microbial and anti-fungal substances to protect the moist, warm eggs. The fluid probably looked hardly at all like milk; but it already contained some of the characteristic milk constituents, such as the product of the ancestral gene to α-lactalbumin and perhaps the casein proteins or at least the ancestors/precursors of those genes. It may have contained little if any lipid (fat); but then some modern milks contain almost no fat (e.g., rhinoceros). Did it contain lactose or other oligosaccharides found in modern milks? Possibly, even probably. But its primary function most likely was to support and protect fluid balance, not to provide metabolizable energy.

In the next chapter we discuss a few of the novel molecules in milk, their probable origins, and how knowledge of the evolution and function of these molecules further clarifies the evolution of milk from a "scarcely nutritious fluid."

Chapter 3

The Molecules of Milk

Milk does not fossilize. We cannot study ancient milk directly. But modern milk contains clues to its evolutionary history in the unique molecules that have arisen to form milk and support lactation. It is in the molecules and DNA where we find evidence for the ancient origins of lactation. The molecular evidence from milk proteins indicate that some form of complex lactation was in existence at least as long ago as a quarter of a billion years. Exactly what that milk was composed of and what were its functions we can only hypothesize; but we know that it contained certain proteins. For example, all milks, whether from a monotreme, marsupial, or placental mammal, contain casein proteins and lactalbumin. Casein proteins are important for their ability to bind amorphous calcium phosphate; most of the calcium in milk is bound into casein micelles. Lactalbumin is essential for producing lactose, the predominant sugar found in milks from most (but not all) mammalian species. In this chapter we focus on a few of the novel molecules, found only in mammals, that are central to lactation: the casein proteins, some of the whey proteins such as α-lactalbumin, and the sugar lactose.

A quick note on terminology. In this and subsequent chapters we will be referring sometimes to a gene and sometimes to the gene product. In some instances we abbreviate the name of the gene and/or its product. When we are talking about the gene the abbreviation will be in italics. For example, prolactin the peptide is abbreviated as PRL and thus the gene for prolactin will be abbreviated as *PRL*.

Milk versus Yolk

Oviparous vertebrates provide most nutrients to their developing embryos by storing them in the yolk of the egg. The vitellogenin genes code for proteins vital to that process. Their nutritional function is similar to that of the caseins. Vitellogenin proteins are calcium-binding proteins; they also serve as precursors for most of the lipoproteins and phosphoproteins in the yolk. Vitellogenin genes are ancient; they are found in egg-laying vertebrates and most invertebrates. There were two ancestral vitellogenin genes in the common ancestor of the amphibians and the amniotes, designated *VIT1* and *VITanc* (Brawand et al., 2008). A duplication of *VITanc* occurred in stem amniotes, producing *VIT2* and *VIT3*; all three *VIT* genes are present and fully expressed in birds (Hillier et al., 2004; Brawand et al., 2008). Remnants of the coding sequences for all three genes have been found in the marsupial gray short-tailed opossum, and pseudogenes related to *VIT1* and *VIT3* have been found in placental mammals (e.g., human, dog, and armadillo). Monotremes retain one functional vitellogenin gene (Brawand et al., 2008); of course, monotremes lay eggs, albeit small ones with relatively little yolk compared with birds.

In the mammalian lineage, milk has replaced yolk as the main source of nutrients for the developing offspring and casein proteins have functionally replaced the vitellogenin genes. This replacement was gradual, and both casein and vitellogenin proteins existed in the early pseudo-lactating synapsids. As evidenced by the continued existence of a vitellogenin gene in all monotremes, this coexistence continued into species with true lactation. The evolution of the nipple (absent in monotremes), which enabled the complex lactation of the marsupials, and the placenta (rudimentary in marsupials and complex in eutherians) allowed the final complete loss of vitellogenin genes in therian mammals (marsupial and placental mammals).

Milk Proteins

"Little Miss Muffet, sat on a tuffet, eating her curds and whey." So begins a well-known, old nursery rhyme. Relatively few people these days may know what all the words mean. For the purposes of this book it matters little whether the reader knows that a tuffet is a footstool covered in fabric so that none of its legs are visible. But curds and whey are important aspects of milk (box 3.1). Milk proteins are typically divided into the casein fraction and the whey protein fraction. Casein proteins form the curds. The word casein is derived from the Latin word for cheese (*caseus*), appropriately, as the processes in making cheese utilize substances (e.g., rennet) to precipitate and coagulate the casein proteins to form an initial curd (box 3.1). Whey is the liquid drained off of the curds; it contains lactose, water-soluble vitamins, and traces of fat in addition to the whey proteins.

Caseins

Casein proteins are usually the predominant proteins in milk. In cow milk, caseins comprise 80% of the protein fraction. The casein protein family in eutherian mammals consists of at least four proteins: αS1-casein, αS2-casein, β-casein, and κ-casein, often with multiple variants. Monotremes and marsupials have only one α-casein; but in monotremes there appears to have been a duplication of the β-casein gene (Lefèvre et al., 2009). All mammals have a single κ-casein gene. The origin of all forms of casein proteins predates the divergence of therian mammals (the metatherians/marsupials and the eutherians/placental mammals) from the prototherian mammals (monotremes) which occurred about 200 million years ago. The α- and β-caseins share similarities that suggest a closer evolutionary relationship to each other than either has to κ-caseins (considered in more detail below). Among the α- and β-caseins, the molecular evidence suggests that the gene duplication and subsequent divergence of αS1- and αS2-casein occurred soon after the divergence of the eutherians from the marsupials, and the duplication and divergence of the two β-caseins genes in the monotremes occurred later, but still before the divergence of the platypus and echidna lineages (figure 3.1).

The casein proteins differ in function. The α- and β-caseins have an affinity for calcium ions while κ-caseins do not. The α- and β-caseins bind amorphous calcium phosphate (box 3.2). This is a vitally important function of the casein proteins. They are the delivery mechanism for calcium and phosphate to the neonate and also serve to protect the mammary gland from tissue calcification. In the absence of casein proteins, amorphous calcium phosphate in milk (in concentrations that would be nutritionally relevant) would form a crytstalline precipitate. Not only would the calcium phosphate not be able to be transferred to the neonate through suckling, but the precipitate would lead to calcification of mammary tissue, endangering maternal health and harming future lactation efforts.

Caseins sequester calcium phosphate by forming micelles (box 3.3) with amorphous calcium phosphate. These are not the same as the fat micelles of milk, the spherical aggregations of lipid molecules that also exist in milk (discussed in chapter 7), but they have similar properties in that they are constructed from molecules with hydrophobic (α- and β-caseins) and hydrophilic (κ-casein) properties. These small spherical micelles are the reason why milk looks white, as they are the appropriate size to scatter visible light.

BOX 3.1

Curds and whey

If you have ever eaten cottage cheese then you have eaten curds and whey. Modern production methods have reduced the amount of whey (since many whey proteins have a bad taste), so cottage cheese these days is much more curds than whey; but both milk fractions are present. Curds and whey are important food components, especially for making cheese. More importantly for this book, they represent milk fractions with different functions for the neonate.

There is a similarity in the initial process for cheese making and the digestion of milk by a mammalian neonate; both start with the precipitation and coagulation of the casein proteins to form a curd. In most cheese making, the substance used is called rennet, a mixture of proteases that modify the structure of κ-casein. Some soft cheeses are made from curds precipitated by the action of acids alone and do not use any proteases. Rennet originally was derived from tissue from the fourth stomach of calves, but now that term refers to any substance that forms a curd from milk. The main active ingredient in the rennet from calf intestine is the enzyme chymosin (also called rennin). Rennet from calf stomach is still used in some cheese making, and there is rennet of plant origin with different active ingredients as well. These days, however, most rennet used by large cheese producers is created by genetically modified microorganisms. These rennets are called fermentation produced chymosin (FPC) and are synthesized by microorganisms that usually have the gene for bovine chymosin inserted into their genome, though some FPCs are produced from the camel chymosin gene. The two major suppliers of FPCs use either the fungus *Aspergillus niger* or the yeast *Kluyveromyces lactis* as the microbial host for the chymosin gene. Thus, most cheese these days is made with and likely contains a substance from a genetically modified organism (GMO) and is therefore technically a GMO food product. It does not, however, contain anything from the genetically modified organisms except the gene product, chymosin.

Chymosin is produced by chief cells in the stomach of all mammalian neonates. It requires a slightly acidic environment to perform its function;   *(continued)*

The micelles need to stay in suspension if they are to be transferred to the neonate and also so that they do not precipitate out and damage the mammary gland. The existence of multiple casein proteins is important for their function in milk. Casein micelles are formed from a combination of the different casein proteins and it is their interaction that keeps these micelles from precipitating.

BOX 3.1 CON'T

a pH of about 6 is optimal. Mammalian neonates secrete less gastric acid and maintain a higher pH stomach content than do adults, presumably an adaptation for milk digestion allowing chymosin to perform its function on mother's milk. Chymosin converts κ-casein to κ-paracasein, which disrupts the stabilizing ability of κ-casein and shrinks casein micelle size and allows micelles to adhere to each other. The result is a cottage cheese–like curd composed of casein micelles, but also trapping much of the fat and even some of the lactose. This is an important first step in making cheese and is an important evolutionary adaptation for mammalian neonates as it results in ingested mother's milk remaining in the stomach and not flowing rapidly into the small intestine. The curdled milk can then be acted upon by other digestive enzymes and by the acidic gastric secretions. If this curdling process does not occur, liquid milk would rapidly flow through the digestive tract with relatively little time for the nutrition to be extracted. Adults produce little or no chymosin, but the curdling action of gastric acid combined with other proteases such as pepsin will also produce a milk curd in the stomach when adults drink milk. The process is not as efficient, however, and more material will stay in the whey and rapidly leave the stomach and enter the small intestine. Neonates are better adapted to get nutrition from milk.

The fact that most cheese manufactured today technically could be labeled a GMO product is little known. The US Food and Drug Administration (FDA) labels FPCs as GRAS (Generally Recognized as Safe). From the FDA perspective, FPCs are simply a more efficient means of producing chymosin but otherwise are identical to chymosin extracted from calf stomachs. Indeed, FPCs are purer, because rennet from calf stomachs contains pepsin and other peptides besides chymosin. For consumers who wish to avoid all GMO-type products, there are some cheeses that are curdled using acids alone (e.g., ricotta, mascarpone, and paneer). There are also cheeses that use either plant rennet, rennets naturally produced by microorganisms, as well as ones still produced from calf rennet. Certified organic cheese is always an option for those who wish not to purchase GMO products. In Europe and the United States any cheese certified as organic cannot, by law, contain FPCs.

The α- and β-caseins are termed calcium-sensitive, as they are subject to precipitation due to interaction with calcium phosphate; κ-casein protects the other casein proteins from calcium ion–induced precipitation (Holt et al., 2013). The existence of any of the α- or β-caseins results in micelles that bind the amorphous calcium phosphate. However, without κ-casein those casein–calcium phosphate complexes would not stay suspended in

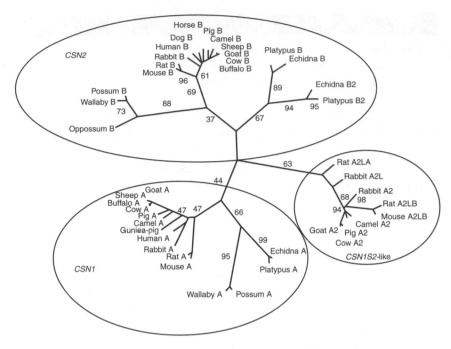

Figure 3.1. Casein molecule phylogeny. From Lefèvre et al., 2009. Reproduced with permission.

BOX 3.2

Amorphous calcium phosphate

Amorphous calcium phosphate (ACP) is found in a number of biofluids and tissues, including saliva, milk, dental enamel, and bone. ACP is important in the mineralization of tissue such as bone and tooth enamel, but also presents a danger to tissue that should not be mineralized. The fundamental unit of ACP is a central calcium ion surrounded by 6 phosphate anions which are surrounded by an additional 8 calcium ions (Posner and Betts, 1975). Casein micelles contain phosphorylated residues that cluster together. These phosphate centers appear to be crucial for binding nanoclusters of ACP. The α- and β-caseins contain phosphate centers while κ-casein does not. A typical cow milk casein micelle will contain about 800 ACP nanoclusters.

milk but would form an amorphous precipitate. Adding κ-casein to the casein–calcium phosphate complex results in stable casein micelles that remain in suspension, sequestering the calcium phosphate and preventing tissue calcification.

The casein proteins are present in milk in high concentration and therefore present a potential danger. Amyloid fibrils are complexes of normally soluble proteins that precipitate to form proteinaceous fibers that are resistant to degradation. These can be toxic, and many diseases derive from amyloid fibrils, such as Alzheimer's disease, mad cow disease, and type 2 diabetes (Rambarran and Serpell, 2008). The high concentration of casein proteins in milk raises the danger of fibril formation and amyloidosis. Both αS2- and κ-casein are susceptible to self-aggregation to form fibrils; however, αS1- and β-casein protect against fibril formation (Holt et al., 2013).

The caseins in milk serve multiple adaptive functions. They interact to protect the mammary gland from calcification and amyloidosis, preserving future reproductive health and potential for the mother after lactation. They serve a nutritional function as a delivery system for calcium and phosphate to the neonate, essential for the proper development of bones and teeth. Caseins also supply amino acids after digestion.

Although casein proteins provide amino acids to the neonate after digestion, the amino acid balance of caseins is not ideal. Caseins are deficient in sulfur amino acids, particularly methionine, and are high in proline, a nonessential amino acid. This is explained by their required structure for the important function of forming the casein micelle and binding calcium phosphate (box 3.3). The importance of the calcium-sequestration function of the casein proteins may have been an evolutionary constraint on its nutritional potential (Holt and Carver, 2012). The amino acid composition that serves to deliver calcium phosphate to the neonate while protecting the maternal mammary gland does not allow caseins to be nutritionally balanced proteins.

EVOLUTION OF CASEIN PROTEINS

Casein protein genes derive from duplications of genes that code for secretory calcium-binding phosphoproteins (SSCP gene family). These are unfolded proteins that are associated with mineralized tissue and are involved in processes such as mineralization of tooth enamel (Kawasaki et al., 2011;

BOX 3.3

Casein and casein micelles

Caseins are unfolded proteins with little tertiary structure that contain a high number of proline residues and no disulfide bridges, hence their paucity of sulfur amino acids. Caseins form structures of many casein proteins bound with calcium phosphate nanoclusters into casein micelles. The word "micelle" is a chemical term. It is used to describe the structure that certain very large molecules will form when dispersed in a solvent. Very large molecules are considered to be too large to be truly soluble in water. Instead, these large molecules will form structures that allow them to remain suspended in water as if they were soluble. The dispersion of these large structures in water is known as a colloidal suspension. The structures that allow large molecules to remain colloidally suspended in water are termed micelles. Under an electron microscope, micelles often look like spheres. In the case of casein, the parts of the casein molecules that have an affinity for water (κ-casein) form the outside of the casein micelle. Conversely, the parts of the casein molecule that are repelled by water (α- and β-caseins) form the inner core of the micelle spheres. A casein micelle will contain thousands of casein molecules bound into a stable complex with nanoclusters of amorphous calcium phosphate. A casein micelle is a functional protein aggregate (as opposed to protein aggregates that lead to pathology) with an amorphous calcium phosphate core.

Oftedal, 2012). A comparison of the chromosomal locations and genetic structure of three members of the SSCP gene family (*ODAM, SCPPQ1,* and *FDCSP*) in a lizard and humans, in conjunction with the location of the casein genes in mammals, suggests that *SCPPQ1* and *FDCSP* derive from duplications of *ODAM* (Kawasaki et al., 2011). The calcium sensitive α- and β-casein genes derive from a duplication of *SCPPQ1*, and κ-casein gene derives from a duplication of *FDCSP* (figure 3.2). Thus, all casein genes originate from duplications of *ODAM*.

The duplications of *ODAM* leading first to *SCPPQ1* and then later to *FDCSP* occurred before the synapsid-sauropsid split. The precursors of the casein genes predate synapsids and probably predate any type of lactation-like adaptation. To date, no casein-like genes have been found in any sauropsid. The ancestral casein genes probably arose in a synapsid, though the dating of the gene duplications that led to the caseins is uncertain.

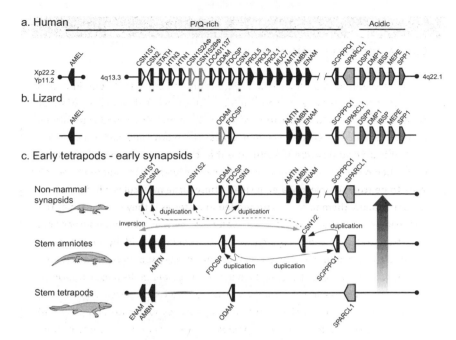

Figure 3.2. SSCP gene family evolution. From Kawasaki et al., 2011; reproduced with permission.

CASEIN AND MILK DIGESTION

The formation of a curd by casein proteins is an important adaptation for the digestion of milk. Liquid generally flows through the digestive tract more quickly than solids, especially when liquid is ingested by itself. If milk was not transformed into a semi-solid by the neonate's digestive tract it would flow quickly through the gut and out the other end, providing little time and opportunity for the extraction of nutrients. However, when milk hits the stomach it is subject to two different chemical processes that reduce the solubility of the casein micelles, causing them to coagulate and precipitate out into what is called a curd (box 3.1).

Milk curds form when milk is exposed to acid, when it is acted upon by chymosin, an enzyme produced by infant stomachs, or by bacterial action (box 3.1). The curdling process is caused by changes in the casein micelles; but curds also contain a sizable proportion of the lactose and fat in milk, as well as water bound up in the curds. The whey proteins, by definition, are not in the curds, and some of the lactose and the water soluble vitamins are also in the whey, not the curds.

Upon ingestion by a neonate, milk is transformed into two fractions: the semi-solid curd and the still-liquid whey. The whey fraction will leave the stomach much sooner than the curd. The curd will be broken down by gastric acid and gastric proteases, releasing nutrients into the small intestine more slowly. Thus, a large milk meal will not result in either a too-fast absorption of nutrients, challenging neonatal metabolism and homeostasis, nor a too-fast exit from the GI tract. The casein proteins provide both a vehicle to provide nutrition and a chemical response to neonatal digestive substances that enhances the effectiveness of milk digestion, delivering a high nutritional return at a metabolically effective rate.

Whey Proteins

Although the caseins make up a large part of the protein fraction, accounting for more than half the protein content in the milk of many mammals, there are many other peptides in milk. These are referred to as whey proteins, as they are not bound up in the curds of milk upon action by chymosin or acid, but are in the whey. The whey proteins comprise the many bioactive peptides in milk that have immune, antimicrobial, and hormonal functions. These peptides may also have some nutritional function as well. Any peptide broken down by neonatal proteases can provide the neonate with amino acids. Because casein proteins are not nutritionally complete, some of the whey proteins must serve a nutritional function, at the least to provide essential sulfur amino acids such as methionine. But many of these whey proteins serve other important functions that have little or nothing to do with nutrition. The whey will exit the stomach relatively rapidly, restricting the time any digestive secretions can act on them, and rapidly enter the small intestine, where they can be absorbed or they may act directly on small intestinal epithelial cells. For example, secretory IgA (sIgA) is the major immunoglobulin in human milk and the milks of other anthropoid primates. Milks of non-primate mammals will often contain high concentrations of IgG and IgM as well, at least in milk produced soon after birth. These molecules transmit the mother's pathogen/immune history to her offspring (discussed in chapter 8).

There are also hormones and growth factors in milk, which may have direct action on the neonate (discussed in chapter 9). Most of these peptides are not exclusive to milk; they can be found in most body fluids and serve functions in many organs and regulatory mechanisms. An extremely interesting and complex attribute of milk proteins is that many of them contain bioactive fragments within the whole. In other words, after the infant's digestive enzymes have acted on the protein by cleaving it into peptide fragments, some of those fragments are thought to have signaling activity.

Two whey proteins (β-lactoglobulin and whey acidic protein [WAP]) are found in the milk of most mammals but, interestingly, are not found in human breast milk. β-lactoglobulin is the main whey protein in most ruminant milks, including dairy cattle. It is found in the milk of monotremes, marsupials, and many placental mammals, implying it is an ancient milk protein. No biological function other than to provide amino acids to the neonate have been demonstrated for β-lactoglobulin, and it is absent from the milk of rodents, camels, llamas, as well as humans (Oftedal, 2012). Indeed, some people develop an allergy to β-lactoglobulin. In a fascinating experiment, a transgenic calf was created that highly expressed a specific microRNA (miRNA 6-4) known to block β-lactoglobulin mRNA translation (see chapter 10 for discussion regarding miRNA). When the calf was hormonally induced to lactate, the milk was free of β-lactoglobulin and had a higher concentration of casein proteins; but otherwise was just like regular cow milk (Jabed et al., 2012). Incidentally, the calf was born without a tail, but was otherwise healthy.

The ancestral gene for the WAP proteins is ancient, as WAP-like genes and proteins are found in vertebrates and invertebrates (Smith, 2011). They are an important constituent of many snake venoms, are found in the egg white of sea turtle eggs, inhibit calcium deposition in the shells of mollusks like abalone, and are assumed to have host-defense function in many crustaceans. All these genes have one or more domains containing eight cysteine residues that form four disulfide bonds. The known functions of these proteins includes antimicrobial action (though the mechanism is not known), anti-protease action, and regulation of calcium deposition.

WAP is a major constituent of rodent whey proteins, but it appears absent in ruminants (e.g., it is not found in cow, sheep, or goat) where the *WAP* gene has been shown to be a pseudogene and not expressed (Hajjoubi et al., 2006). It is present in both of the monotreme genera, though the structure differs from placental mammal *WAP* (Sharp et al., 2007); not surprising considering the 200 million years of separate evolution between monotremes and placental mammals. It is expressed in tammar wallaby milk during mid-lactation, with a structurally related peptide (WFDC2) expressed during early lactation. Other species with milk WAP include camels and pigs. As mentioned previously, human breast milk does not contain WAP.

The function of WAP in milk is not understood. Antimicrobial or anti-protease activity has not been demonstrated. The suggestion that WAP is important for the development of the mammary gland founders on the observation that the mammary glands of *WAP*-null mice develop normally with normal milk production, at least through the first half of gestation. Growth of pups of *WAP*-null mice does falter in the second half of lactation, suggesting that WAP in mice has some important effect on mammary gland function, milk composition, or pup development. The answer may be nutritional, as WAP has the highest sulfur amino acid composition of any of the major milk proteins. As mentioned above, caseins, due to their required unfolded structure, are low in sulfur amino acids. It is possible that WAP has lost any antimicrobial, protease inhibiting, cellular development functions in mammal milks and has been retained primarily as a source of sulfur amino acids (Oftedal, 2012).

The whey protein we discuss in more detail in this chapter, α-lactalbumin, has a function unique for milk (producing lactose), and the evolution of α-lactalbumin was a seminal development necessary for the evolution of milk from the proto-lacteal secretions of early synapsids. The gene for α-lactalbumin is only expressed in the mammary gland, making it a truly milk-specific peptide. The α-lactalbumin gene derives from a duplication of the lysozyme c gene. We start our consideration of α-lactalbumin by exploring lysozyme c gene family evolution.

Lysozyme Gene Family Evolution

Lysozyme c is an ancient bacteriolytic enzyme that is ubiquitous in nature, being found in vertebrates and invertebrates (the c stands for chicken-type, since it was first found in chicken egg white). In mammals it is found in body fluids (e.g., blood, tears, and milk); in birds it can be found in high concentrations in eggs (Irwin et al., 2011). Its main function is to lyse bacteria cell walls, and thus lysozyme is part of the innate immune system. It is one of the major defenses against bacterial infection in the eye. Its presence in milk serves to protect the infant and the mammary gland from some bacterial infections.

The bovine genome contains a large number of apparently expressed duplicates of the lysozyme c gene, at least 12 (Irwin et al., 2011). Many other mammalian herbivores also show multiple copies of the lysozyme c gene, including many herbivorous rodents and the elephant. Many of these genes are expressed in the intestinal tract and appear to act as digestive enzymes (Irwin, 1995;

Irwin et al., 2011). The bacterial cell wall lysing function of lysozyme c has been co-opted to digest the copious amounts of bacteria that assist fiber digestion in mammalian herbivores, recycling some of the nutrients lost in fermentation. An immune function molecule has spawned digestive molecules.

Initially, the vertebrate lysozyme gene family was thought to consist of three genes: lysozyme c, α-lactalbumin, and calcium-binding lysozyme. Lysozyme c is ubiquitous among vertebrates. Calcium-binding lysozyme is rare, but taxonomically widespread, being found in a few mammals (e.g., in carnivores such as dog, cat, and seal, as well as phylogenetically quite different species, such as the horse and echidna), as well as in the eggs of pigeons and the stomach of the foregut-fermenting bird, the hoatzin. It is tempting to wonder if calcium-binding lysozyme is performing some kind of digestive function in the hoatzin, or possibly is regulating the stomach microbiome of this foregut-fermenting bird. Only mammals have the gene for α-lactalbumin. A functional gene for α-lactalbumin is found in almost all mammals, where it has a fundamental role in lactose synthesis. The exceptions are sea lions, fur seals, and walruses, which have a pseudogene and produce milk with no lactose (see chapter 7).

Recently, broad searches of vertebrate genomes for lysozyme-like genes have identified many more potential members of this family. As many as eight lysozyme gene family members may have existed in the ancestral mammal, and many are still expressed in the different mammalian lineages. Four lysozyme-like genes have been discovered to be expressed in testes, and two of them have been shown to have a reproductive function in mice (Irwin et al., 2011). Interestingly, the platypus has a lysozyme-like gene that appears to be unique among mammals; at least it has not been found in any placental mammal or marsupial to date. Either it was lost in the lineage leading to marsupial and placental mammals, or a new gene duplication event has occurred in the monotreme lineage.

The evidence suggests that the duplication of the ancestral lysozyme gene that led to α-lactalbumin occurred before the synapsid-sauropsid split, and these two genes are the ancestral genes for lysozyme c and calcium-binding lysozyme. One current hypothesis is that the ancestral gene for calcium-binding lysozyme also underwent a duplication event, and the duplicate evolved into α-lactalbumin. Calcium-binding lysozyme has been lost from most mammal genomes, but the functional gene exists in phylogenetically diverse lineages (e.g., echidna, horse, dog, and sloth) and a pseudogene exists in taxa as diverse as elephants and primates (Irwin et al., 2011), implying it originally had a widespread distribution in placental mammals. Either the duplication event that led to α-lactalbumin occurred after the synapsid-sauropsid split, or the ancestral α-lactalbumin gene was lost in the sauropsids (Prager and Wilson, 1988; Irwin et al., 2011; Oftedal, 2012).

Some researchers have questioned whether calcium-binding lysozyme in birds and mammals are actually descended from the same gene duplication, and suggest they may derive from independent duplications of lysozyme c (Irwin et al., 2011). The evidence is strong that the genes for mammalian calcium-binding lysozyme and α-lactalbumin are orthologous (derived from the same ancestral gene duplication of lysozyme c), as the dog calcium-binding lysozyme gene and the primate pseudogene are in the same genomic neighborhood of the α-lactalbumin gene (Irwin et al., 2011). If mammalian and bird calcium-binding lysozyme genes are paralogous (independently deriving from lysozyme c), then the α-lactalbumin and mammalian calcium-binding lysozyme genes likely diverged after the synapsid-sauropsid split. However, earlier work suggested a more ancient divergence of α-lactalbumin from lysozyme c of approximately 310 million years ago (Prager and Wilson, 1988), prior to the synapsid-sauropsid split. If so, that would imply that the bird and mammalian calcium-binding lysozyme genes are orthologous, or the ancestral gene of mammalian calcium-binding lysozyme and α-lactalbumin was lost in all the sauropsids. The competing hypotheses and their consequences for the possible age of lactation as an adaptive feature in amniotes are presented in box 3.4.

Regardless of the evolutionary origin of the α-lactalbumin gene, the original function of all these gene products probably was microbial cell lysing. The ancestor to the α-lactalbumin gene evolved over time to perform a unique enzymatic function, allowing the evolution of what is now a uniquely mammalian feature: lactose production.

Lactalbumin

The importance of α-lactalbumin to lactation cannot be overemphasized. This novel protein allowed a novel sugar to be formed: lactose. The primary function of α-lactalbumin is to combine with the enzyme β-1-4-galactosyltransferase to form a heterodimer enzyme called lactose synthase. This enzyme has a high affinity for glucose, much higher than β-1-4-galactosyltransferase has on its own, and catalyzes the reaction: UDP-galactose + glucose = lactose.

The result is a high rate of synthesis of the disaccharide lactose (glucose-galactose) which otherwise is rare if not absent in nature. Mammals appear to be the only organisms able to produce lactose, and lactose is only produced in the mammary gland.

Although in modern mammals α-lactalbumin gene expression is apparently restricted to the mammary gland, there are related genes that are expressed in other tissues, as noted above. In the ancestral condition, the ancestral gene for α-lactalbumin was probably widely expressed. Of course, that ancestral gene probably had a function similar to calcium-binding lysozyme or, more accurately, the function of the ancestral calcium-binding lysozyme peptide. In the beginning it had to have had identical function, as it was a duplication of that gene. Over generations the two genes diverged in sequence and hence gained the ability to diverge in function. But originally, the ancestral lactalbumin gene may have had bactericidal activity, and its main function may have been in immunoprotection against microbes.

Does lactalbumin retain any bactericidal function? In its normal folded configuration the answer appears to be no. However, when human milk is treated with acid in order to precipitate out the casein proteins, a partially unfolded configuration of lactalbumin, stabilized by oleic acid, is formed and appears to get trapped in the casein fraction, not the whey. This peptide, called human α-lactalbumin made lethal to tumor cells (HAMLET), exhibits cytotoxic properties (Mossberg et al., 2010). An exciting find of a potential anti-tumor molecule; but does HAMLET have biological function in a mother–nursing baby dyad? Does it even exist in nature?

HAMLET was discovered by chance when scientists were trying to examine characteristics of casein proteins. They used acid, not chymosin, to precipitate the caseins, and that acid treatment resulted in the modification of the α-lactalbumin protein. This implies that HAMLET probably would be formed in the low pH environment of the adult stomach after milk ingestion. Neonates produce less gastric acid and hence have a higher stomach pH, suitable for chymosin activity, as mentioned above. It is not clear if HAMLET would be produced in the stomachs of neonates. Antimicrobial properties of α-lactalbumin under normal circumstances of an offspring nursing from its mother through this mechanism are speculative at this time. Still, some evidence for an antimicrobial function for α-lactalbumin exists. In vitro it does act to reduce the growth of some microbes. After

BOX 3.4

When did the ability to produce lactose first occur?

Many numbers for the age of lactation and origin of milk have been suggested. The difficulty, of course, is that milk and the mammary gland do not fossilize. The dates for the origin of lactation and of milk (they perhaps can be different dates, depending on definitions) must be extrapolated from phylogeny and examination of the genes that appear fundamental to lactation and milk. We know that about 200 million years ago mammals came into existence, and that the existing branches of mammals that derive from those creatures all lactate and produce what we call milk. Since the living monotremes and metatherians (marsupial and placental mammals) diverged about 160 million years ago, modern milk existed since at least then and probably back to 200 million years ago. Some authors have argued that the origin of lactation might even extend to further than 300 million years ago. One piece of evidence in favor of such an ancient origin is the 310-million-year estimated date of when the gene duplication event occurred that led to modern lysozyme c on one branch and α-lactalbumin on the other. The argument is that α-lactalbumin, with its ability to catalyze the production of lactose, is a fundamental precursor to lactation.

We agree that the ability to produce lactose is a crucial element in the evolution of a substance that would be considered milk, though the existence of mammals (e.g., sea lions and walruses) without a functional α-lactalbumin gene who therefore produce a lactose-free milk emphasizes that lactose, lactation, and milk are not synonymous. We argue that for a substance to be called milk it needs to contain casein micelles and fat globules contained within the fat globule membrane, but it does not need to contain lactose. However, lactose may have been crucial for the successful secretion of a watery substance by ancient synapsids before the evolution of casein micelles and the fat globule membrane. Lactation, defined as the secretion by reproductive females of a fluid beneficial to offspring, came first. The fluid evolved to become milk.

If the ancestral gene for α-lactalbumin originated 310 million years ago, does that truly imply that lactation was either already in existence or soon to come into being? Not really. That gene was a duplicate of an existing gene that   *(continued)*

digestion some fragments of the α-lactalbumin molecule appear to have antimicrobial action; not surprising considering its evolutionary lineage.

This illustrates a fascinating aspect of many milk proteins. When whole they may have a certain function, or even no obvious function; but after proteases break the milk protein into smaller peptides, many of these seem to

BOX 3.4 CON'T

most likely produced a bactericidal peptide that damaged bacterial cell walls, similar to lysozyme c function today. At some point it mutated to gain its enzymatic ability to catalyze the formation of lactose, but we have no real idea of how soon that happened. It certainly had occurred before the ancestor of all living mammals, so probably more than 200 million years ago. But what the function of that gene product was 300 million years ago is unclear, and probably unknowable. The gene duplication event may have been a necessary precursor to the evolution of lactation, but it does not definitively date the origin of lactation.

Further complicating the dating of the existence of α-lactalbumin and hence the origin of lactose production is that α-lactalbumin may derive from a subsequent gene duplication, with the surviving gene lineages representing α-lactalbumin (only found in mammals) and calcium-binding lysozyme (found in birds and apparently in some mammal taxa, both as a functional gene and as a pseudogene). The most parsimonious phylogeny has the original lysozyme duplication event around 310 million years ago giving rise to the lysozyme c lineage and a lineage that leads to both calcium-binding lysozyme and to α-lactalbumin. The subsequent gene duplication of the ancestral gene to calcium–binding lysozyme and α-lactalbumin would have occurred in synapsids after the split between the sauropsid and synapsid lineages. Again, neither of those gene products was likely to have any lactose-producing ability. At some point the gene product of the gene that was ancestral to α-lactalbumin actually became sufficiently similar in structure to catalyze lactose formation, but whether that was a few million years after the gene duplication or several tens of million years is unknown.

Lactation, defined as a maternal adaptation of providing a nurturing fluid for offspring, first occurred more than 200 million years ago and possibly more than 300 million years ago. That secretion may have always contained sugars (oligosaccharides), but it could not have included lactose in any form until possibly 310 million years ago but more probably significantly less long ago than that, as the ancestor gene for α-lactalbumin accumulated mutations that eventually gave it the ability to catalyze lactose synthesis. The first existence of what we would truly call milk, with its casein micelles and fat globules contained in a special membrane, probably occurred sometime between 200 and 300 million years ago.

have the potential for biological function. Milk contains potential bioactive molecules that require the action of the neonate's digestive tract to activate function.

At some point in time α-lactalbumin gained the ability to act as a coenzyme with β-1-4-galactosyltransferase and produce lactose, a novel sugar. It is not clear whether the early proto milk contained free lactose or oligosaccharides with lactose residues. The milks of monotremes and marsupials generally contain more oligosaccharides than free lactose, suggesting the latter. Many eutherian mammals contain more lactose than other, larger oligosaccharides. Indeed, lactose is the second most common milk constituent (after water) in the milks of primates and perissodactyls. Interestingly, human milk contains the highest diversity of milk oligosaccharides among placental mammals (discussed in chapter 12). Human milk oligosaccharides always contain lactose on the reducing end. We discuss the functions of lactose in modern milks and our ideas about its function in ancient proto milk in the next section.

Lactose

Lactose was discovered in milk in 1619 by the Italian physician Fabrizio Bartoletti (1576–1630) and identified as a sugar in 1780 by the Swedish Pomeranian chemist Carl Wilhelm Scheele (box 3.5). Lactose in milk serves several functions. It is the main nutritional sugar in many placental mammal milks; less so in monotreme or marsupial milks where larger oligosaccharides are predominant. All mammalian milks contain oligosaccharides. Milk oligosaccharides typically have a lactose molecule on the reducing end, so lactose production is an integral part of milk oligosaccharide production. Whether early proto milk contained free lactose or only lactose attached to larger oligosaccharides is uncertain. Based on the predominance of larger oligosaccharides in monotreme and marsupial milks, some authors have suggested that early milk carbohydrate was primarily oligosaccharides, with only trace amounts of lactose (Urashima et al., 2012). In placental mammals, lactose appears to provide metabolizable energy in milk and the other milk oligosaccharides perform other functions, generally assumed to be prebiotic and/or antibiotic. The high concentration of milk oligosaccharides in monotreme and marsupial milks implies that these neonates possess the ability to digest and metabolize milk oligosaccharides.

In placental mammals, lactose acts to create an osmotic gradient into the Golgi apparatus of mammary epithelial cells, drawing water into the Golgi lumen and eventually into milk along with the other milk constituents. This consequence of lactose production will be considered in more detail in chapter 7 to explain why high-sugar milks are always high-water milks. Of course almost any sugar would produce an osmotic gradient drawing water into the mammary gland and thus into milk. Other milk oligosaccharides and even monosaccharides, such as glucose, would create an osmotic gradient. If food for a neonate was formulated rather than evolved, glucose might be considered to be a more appropriate carbohydrate energy source, requiring no digestion and being immediately metabolizable. The advantage of lactose as a disaccharide is that it delivers twice the potential metabolizable energy in milk for the same amount of water. In other words, a 7% glucose milk and a 7% lactose milk have the same osmotic effect, but the 7% lactose milk has a galactose molecule for every glucose molecule, effectively doubling the carbohydrate energy. Of course, larger oligosaccharides might offer even greater gains, in theory. In practice, the difficulty in digesting larger oligosaccharides and other aspects of their composition and structure reduce any advantage they would have over lactose.

The original functions of oligosaccharides, whether containing lactose or not, in ancient proto milk is uncertain. We propose that one function was to draw water into the gland to create a watery secretion. We hypothesize that oligosaccharides originally did not perform a nutritional function. They created an osmotic gradient. The advantage of oligosaccharides rather than simple sugar molecules, such as glucose, was that microbes could not easily use them for metabolism. In following the theory of Oftedal (2002b; 2012), eggs were being bathed in a watery fluid secreted from a ventral gland of an endothermic synapsid. Warm sugar water spread on eggs would have been an invitation for fungal, bacterial, and other microbial growth. Antimicrobial substances would have been favored for addition to the fluid; but a simple and effective adaptation was to use sugars that are difficult for microbes to digest and metabolize.

An advantage of the disaccharide lactose over larger oligosaccharides is that it is cheaper to produce (takes fewer sugar molecules) and it was novel. Even today, many yeasts and other microbes lack the ability to digest lactose. Among vertebrates, only mammals produce lactase, the lactose digesting enzyme. Of course, over the hundreds of millions of years since lactose was first produced, many microbes evolved lactose-digesting ability. But initially, lactose would have been the perfect sugar to draw water into the gland, producing a watery secretion to bathe the eggs that was carbohydrate-rich but resistant to microbial attack.

Could lactose have had function outside of milk? Oligosaccharides cover cell membranes and are important in cell-to-cell communication and as adhesion sites for pathogenic microbes to gain entry to the cell. The existence of α-lactalbumin allows the production of lactose and its incorporation into oligosaccharides, producing novel oligosaccharides that are found only in

BOX 3.5

Hard-luck Scheele

Carl Wilhelm Scheele (1742–1786) was born in Pomerania, an area on the Baltic coast of Germany and Poland which at the time was ruled by the Swedish empire. Between the ages of 14 and 22 Scheele served as an apprentice pharmacist in Gothenburg, Sweden. His interests in chemical experimentation began during his apprenticeship and continued through his life. In 1770 he became the director of the laboratory of the pharmaceutical company of Christian Ludvig Lokk in Uppsala. Uppsala was an important academic center in Sweden at that time, and here Scheele began his life-long association with the eminent professor Torbern Bergman (1735–1784). In 1775 Scheele was elected to the Royal Swedish Academy of Sciences. In that year he moved to Köping, Sweden, where he lived out the rest of his life at his own pharmacy, but mostly doing his chemical experiments.

Scheele has been credited with discovering a large number of new-at-the-time substances: for example, chlorine (which he called dephlogisticated marine acid), hydrogen fluoride, silicon fluoride, hydrogen sulfide, and hydrogen cyanide. He isolated and characterized a number of substances, such as benzoic acid, citric acid, lactic acid, and tartaric acid, as well as glycerol and the pigment copper arsenite (Scheele's green). He is thought to have been the first to discover barium, manganese, molybdenum, and tungsten. In 1780 he determined that the substance isolated from milk by Fabrizio Bartoletti, lactose, was a sugar.

Scheele was given the sobriquet "hard-luck Scheele" by Isaac Asimov due to the fact that despite now being recognized as being the first to discover many chemical substances, during his time his contributions were known only to a few. Priestley and Lavoisier were appropriately feted at the time for their work on oxygen; but Scheele's experiments had preceded them. However, his book detailing his work was published several years after Priestley's work was   *(continued)*

milk, as best we can tell from the literature. We have not been able to determine if any cell-surface oligosaccharides in mammals contain lactose, though the generally accepted view is that α-lactalbumin is only expressed in the lactating mammary gland, which would suggest not. Lactose is found on the reducing end of milk oligosaccharides, which generally would not be considered an attachment site for microbes. Therefore, it is probably more science fiction than science to suggest that expression of the ancient α-lactalbumin molecule in multiple tissues may have had an adaptive func-

BOX 3.5 CON'T

published. A letter he wrote to Lavoisier, which describes his findings, was sent to Lavoisier shortly before Lavoisier met with Priestley and learned of Priestley's experiments (Lavoisier did not answer this letter).

This does not mean that Priestley or Lavoisier stole from Scheele. Both those scientists were working independently on properties of air. Scheele may have performed experiments to separate oxygen first (heating manganese oxide and mercuric oxide to liberate the oxygen), but Priestley independently demonstrated that air contained an element that sustained fire (and life), and Priestley even had shown that plants released this element. Lavoisier wrote that he, Priestley, and Scheele all discovered oxygen independently at about the same time (a claim that Priestley vigorously disputed, at least as far as Lavoisier's claim to be a discoverer of oxygen). Scheele referred to the substance he had found as "fire air." Priestley called it "dephlogisticated air." It was Lavoisier who concluded it was a new element and termed it oxygen.

Scheele was described as a self-effacing man and he was slow to publish. He was a careful laboratory scientist who performed his experiments as much for his own understanding as for any external credit. Priestley recognized the importance of his own discovery and quickly published, but he never came around to the correct interpretation of his results and remained convinced of the phlogiston theory until his death. Lavoisier was the last of these to actually perform experiments that captured oxygen, and he only repeated what Priestley and Scheele had already done (and told him about). Still, Lavoisier was the closest to understanding the true nature of oxygen. All three contributed to science.

Being a chemist in those days unfortunately was not conducive to health. Scheele had a bad habit of sniffing and tasting any new substances he created. Cumulative exposure to arsenic, mercury, lead, their compounds, and perhaps hydrofluoric acid, which he had discovered, took their toll on Scheele, who died at the early age of 43 at his home in Köping. Doctors said that he died of mercury poisoning.

tion in producing novel cell-surface oligosaccharides that were invisible to pathogens. As best we can tell, that is not the case today. Even if it did occur in the past, the arms race between vertebrates and microbes usually is biased in favor of the microbes. Any short-term advantage in producing cell-surface oligosaccharides that existing microbial strains could not attach to would in the longer term just provide an adaptive niche for synapsid-specific microbes that evolved an ability to recognize and use them. The importance of cell-surface oligosaccharides in cell-to-cell communication also argues that cell-surface oligosaccharides may be somewhat conservative, as changes that reduced microbial attachment might also negatively affect cell-to-cell communication. It would appear that the ability to produce the novel sugar lactose may have been a lactation-specific adaptive event.

More detailed discussion of the functions of milk oligosaccharides and the importance of lactose, by itself and in large oligosaccharides, can be found in chapters 7, 8, and 12. Glycobiology is a complex subject. Neither of us is a glycobiologist, so we approach milk oligosaccharides with great caution, and even so we can be accused of temerity in proposing adaptive functions for molecules for which we have little understanding. However, oligosaccharides are fundamental to the evolution of milk and lactation, so we have no choice but to be brave and propose hypotheses to be considered, and present what supporting evidence exists.

A Scarcely Nutritious Fluid

We end the third chapter by returning to the concerns of Mivart as outlined in the beginning of chapter 2 regarding the evolution of milk. What was that scarcely nutritious fluid from which milk evolved, what were its adaptive functions, and how could it have evolved to become milk?

Mivart assumed that the original proto-lacteal secretion was similar in composition to sweat; mainly water and some electrolytes. Not an unreasonable assumption, especially considering the technology and scientific knowledge at the time, and possibly correct. However, we suggest a better analogy might be saliva, a fluid containing many secreted proteins as well as minerals, including amorphous calcium phosphate. Still scarcely nutritious, but quite biologically active. Calcium-binding proteins would likely have been present, to protect the proto mammary gland from calcification. If the ideas of Kawasaki and colleagues (2011) on the origins of the casein genes hold up, one of those calcium-binding proteins was the ancestor of the α- and β-caseins, and another the ancestor of κ-casein.

We accept as most probable that the primary adaptive function of this proto milk was providing moisture to eggs to prevent desiccation, as per Oftedal (2002b; 2012). Other authors have suggested that the original proto lacteal secretions were involved in the innate immune system related to the inflammatory response and antimicrobial actions (Vorbach et al., 2006). The two hypotheses are not exclusive and could be considered complementary. A logical consequence of Oftedal's hypothesis regarding the function of proto milk is that both the eggs and the mother's skin would have been kept moist and warm (synapsids likely were endothermic). Moist, warm skin and moist, warm eggshell both would be fertile places for microbial growth. Antimicrobial substances to protect both the mother's skin as well as the eggs from bacterial and fungal attack would have been strongly selected for inclusion into the proto milk. These substances likely included lysozyme c, ancestral WAP, and the product of the duplication of the calcium-binding lysozyme gene that was ancestral to α-lactalbumin, which likely had anti-biotic properties before its continued evolution into a coenzyme to promote lactose formation.

There probably were many other substances in proto milk that could have been utilized by the developing embryo, assuming they could have been absorbed through the parchment shell of the egg. Certainly calcium phosphate absorbed through the shell could have been used for bone mineralization of the developing embryo. The casein micelles that bind calcium phosphate in milk probably would have been too large to enter the egg; but simpler, smaller complexes of calcium-binding proteins and calcium phosphate could have managed. The concentrations of these constituents were probably low in proto milk; after all, the primary function was water balance and keeping the eggs from desiccating. We hypothesize that parental care after hatching was a necessary (but not sufficient) characteristic for the proto milk to experience selective pressure to increase and concentrate the nutrients it contained; specifically, that hatchlings ingested the proto-lacteal secretions. It is reasonable to suggest that early female synapsids that exhibited parental care toward their eggs also would have exhibited parental care toward hatchlings, and there is some evidence in support of parental care of young in later synapsids, as described in chapter 2. All that would be required would be the continuation of maternal secretion of proto milk for some time after hatching and for hatchlings to lick the proto milk off their mother's skin. The original main adaptive value could still be water balance. The proto milk may have been "scarcely nutritious" but it contained plenty of water, and water may have been a limiting nutrient for hatchlings as well as for eggs. The ability to lactate may have enabled early synapsid hatchlings to survive in the increasing areas of arid environments, moderating the stress of water balance on them, though possibly increasing maternal water stress. Adult animals are larger and generally more able to cope with water and food stress. Many seasonally reproductive mammals lactate during the "bad" months, when food and water is scarce. Their offspring are fed milk during the time period where food and water would be hard for them to acquire (see chapter 7).

Mothers are certainly put under stress by this strategy; but they have more nutrient stores on their bodies and are more competent at obtaining what food and water is available. The ability to provide their eggs and then their hatchlings with a water source may have been a strong adaptive feature that allowed proto-lactating synapsids to survive and reproduce in the arid landscapes surrounding the shrinking rain forests.

Proteins and minerals in proto milk would have provided nutritional benefit to the hatchlings. Antimicrobial molecules, including immune function molecules, could have also provided benefits. However, in our opinion it was the evolution of an α-lactalbumin-like molecule and the ability to produce the novel sugar lactose that was a vital step from proto milk toward milk. Now proto milk potentially became a carbohydrate-rich secretion. But it was still awaiting one last vital evolutionary event: the evolution of an enzyme to digest lactose.

Once hatchlings began ingesting proto milk, selective pressures to increase the nutrient content of proto milk would have been strong. Casein micelles, which may not serve an adaptive purpose when secreted onto eggs, would have a potentially profound adaptive purpose when ingested by hatchlings, increasing both protein and calcium-phosphate concentrations. Lipids secreted into the milk would provide energy and potentially also important long-chain polyunsaturated fatty acids that are important in membranes. Vitamins secreted into the milk would further increase its adaptive significance (though perhaps not vitamin D; see chapter 5). The final evolutionary event that sealed the deal for lactation and milk was the evolution of lactase, a member of the β-galactosidase enzyme family, which enabled the digestion of lactose into its constituent sugars, glucose and galactose. The glucose could be easily used in metabolism by synapsid hatchlings. The galactose might have presented greater difficulty, but either enzymes to convert galactose to glucose already existed or were soon evolved. Galactose can be converted to glucose by all mammals, although there are genetic conditions that prevent its conversion and result in potentially fatal galactosemia in human infants.

There are, of course, other molecules important in lactation and milk that are not novel, that came into being long before the synapsid lineage came into existence. These molecules are found in most if not all vertebrates, and some of them can be found in invertebrates. These ancient molecules were co-opted to serve lactation, evolving new function. We will discuss two examples (oxytocin and prolactin) in chapter 4.

Prolactin and Oxytocin

A hallmark of an evolved system is the co-option and repurposing of existing structures and molecules. Lactation evolved from existing structures and its regulation uses ancient regulatory molecules that have evolved multiple functions in mammals and other taxa. Prolactin and oxytocin are crucial molecules involved in the process of lactation. They are also ancient molecules linked to many other diverse functions in both the periphery and the brain in a wide range of taxa, including invertebrates as well as vertebrates. The ancestral molecules of oxytocin and prolactin both came into being more than half a billion years ago.

We examine these two molecules as examples of the fundamental evolutionary process by which existing gene products are co-opted into new function. Both of these peptides were probably involved in the earliest phases of the evolution of lactation. Oxytocin is involved in the birth process, including egg laying in such diverse taxa as insects, birds, and monotremes. Given our preferred proposed evolutionary origin of lactation as a mechanism to secrete a fluid onto eggs (sensu Oftedal), to co-opt oxytocin into this secretory process cued by egg laying has biological plausibility. Prolactin is linked to parental care behaviors in many taxa, including birds and mammals. It is important in both lactation and in crop milk, the analogous lactation-like strategy of several bird lineages (see chapter 1). Thus, prolactin as a molecule linked to parental care predates the synapsid-sauropsid split. Again, there is biological plausibility for co-option into lactation from the beginning. Prolactin is also involved in aspects of fluid balance in taxa as diverse as fish and mammals, as is oxytocin's sister molecule vasopressin or its equivalents in other taxa. It makes intuitive sense that these molecules would have been co-opted for a parental process that involves secreting a fluid whose original adaptive purpose may have been buffering offspring (eggs and/or hatchlings) from water loss.

In this chapter we will review the functions of oxytocin and prolactin, focusing on their functions in lactation but also exploring their other known functions to investigate why these molecules may have been co-opted into the maternal behavior of lactation.

Oxytocin

In 1955, Vincent du Vigneaud was awarded the Nobel Prize in Chemistry "for his work on biochemically important sulfur compounds, especially for the first synthesis of a polypeptide hormone." That peptide was oxytocin. Oxytocin is an ancient, small (nine-amino-acid) peptide with a wide range of functions in many different organ systems. In the brain it is linked to attachment and affiliation among individuals. Sir Henry Hallett Dale (who shared the Nobel Prize for the discovery of acetylcholine) demonstrated its uterine contraction properties in 1906. In egg-laying species it is important in oviduct contractions; in other words for laying eggs. The name oxytocin comes from the Greek meaning swift birth. It has fundamental importance in milk let-down, allowing milk to be expressed from the nipples. The milk ejection reflex due to oxytocin was described in 1910 (Ott and Scott, 1910) and 1911 (Schafer and Mackenzie, 1911).

Circulating oxytocin is produced in the hypothalamus (mainly by magnocellular neurons of the paraventricular and supraoptic nuclei of the hypothalamus) and transported to the pituitary gland, where it is released into circulation. Oxytocin is also made locally in diverse tissue. For example, heart tissue synthesizes oxytocin, which through paracrine action stimulates the release of atrial natriuretic peptide, providing local signaling important for regulating vascular tone and heart rate (Jankowski et al., 1998; Gutkowska et al., 2000). It is also known to be produced by the ovaries, testes, and lungs, and by the placenta and decidua during pregnancy (Kim et al., 2015). It is not known to be produced by the mammary glands. Oxytocin release by the pituitary in response to nipple stimulation is responsible for milk let-down.

Vasopressin/Oxytocin Gene Family

Oxytocin is an ancient peptide; the ancestral gene is estimated to have come into existence at least 500 million years ago and probably as much as 600 million years ago given the existence of similar peptides in invertebrates. An oxytocin/vasopressin-like gene is found in arthropods, annelids, and mollusks. A gene duplication event occurred before the common ancestor of most vertebrates, resulting in almost every vertebrate having an oxytocin-like peptide with reproductive function and a vasopressin-like peptide with fluid balance/osmoregulatory function. Lampreys are the exception with only

Table 4.1. Known members of the vasotocin superfamily of nonapeptides

Vasotocin (non-mammalian vertebrates—ancestral form)	Cys	Tyr	Ile	Gln	Asn	Cys	Pro	Arg	Gly
Oxytocin (most placental mammals)	Cys	Tyr	Ile	Gln	Asn	Cys	Pro	**Leu***	Gly
Proline-oxytocin (NWM)	Cys	Tyr	Ile	Gln	Asn	Cys	Pro	**Pro**	Gly
Mesotocin (most vertebrates)	Cys	Tyr	Ile	Gln	Asn	Cys	Pro	**Ile**	Gly
Vasopressin (most mammals)	Cys	Tyr	**Phe**	Gln	Asn	Cys	Pro	Arg	Gly
Lypressin (pigs and hippos)	Cys	Tyr	**Phe**	Gln	Asn	Cys	Pro	**Lys**	Gly
Phenypressin (some marsupials)	Cys	**Phe**	**Phe**	Gln	Asn	Cys	Pro	Arg	Gly
Seritocin (frogs)	Cys	Tyr	Ile	Gln	**Ser**	Cys	Pro	**Ile**	Gly
Isotocin (bony fishes)	Cys	Tyr	Ile	**Ser**	Asn	Cys	Pro	**Ile**	Gly
Glumitocin (skates)	Cys	Tyr	Ile	**Ser**	Asn	Cys	Pro	**Gln**	Gly
Diuretic hormone (locust)	Cys	**Leu**	Ile	**Thr**	Asn	Cys	Pro	Arg	Gly
Annetocin (earth worms)	Cys	**Phe**	**Val**	**Arg**	Asn	Cys	Pro	**Thr**	Gly
Lys-connopressin (snails, leeches)	Cys	**Phe**	Ile	**Arg**	Asn	Cys	Pro	**Lys**	Gly
Arg-connopressin (cone snail)	Cys	**Ile**	Ile	**Arg**	Asn	Cys	Pro	Arg	Gly
Cephalotocin (octopus)	Cys	Tyr	**Phe**	**Arg**	Asn	Cys	Pro	**Ile**	Gly
Octopressin (octopus)	Cys	**Phe**	**Trp**	**Thr**	**Ser**	Cys	Pro	**Ile**	Gly

*Amino acids listed in **boldface** indicate changes from the ancestral sequence.

a single member of this gene family (Wallis, 2012). All these peptides have nine amino acids (table 4.1). Some taxa have more than two nonapeptides that derive from the ancestral gene, including the marsupial mammals, which have three members of this gene family (Wallis, 2012). Interestingly, the monotreme and the placental mammals generally only have oxytocin and vasopressin, suggesting an independent duplication event in the metatherian lineage after it diverged from the eutheria. Among invertebrates, the octopus is the only known taxa to have two members of this gene family (Takuwa-Kuroda et al., 2003; table 4.1)

In all mammals the oxytocin/vasopressin genes are located on the same chromosome and relatively close to each other. In marsupials the orientation is tail-to-head, meaning that the genes are read in the same direction; in monotremes and placentals the orientation is tail-to-tail, with the genes read in opposite directions. The tail-to-head orientation is found in the non-mammalian vertebrates (e.g., bony fishes, amphibians, and reptiles), indicating either an independent change in orientation in monotreme and placental mammals or that the tail-to-tail orientation was ancestral in synapsids and was later reversed in marsupials, possibly related to the additional gene duplication.

The structure of oxytocin and vasotocin are similar, but the differences are functionally important. Oxytocin and oxytocin-like molecules have a non-polar, neutral amino acid at position 8, while vasopressin and its relatives have a basic polar positive amino acid (arginine or lysine). All have cysteine at position 1 and 6 and glycine at position 9 (table 4.1). Oxytocin has the amino acid isoleucine at position 3, and that is important for binding with its receptor. However, the armadillo appears to have leucine replacing isoleucine at position 3 (Wallis, 2012). Some, but not all, New World monkeys have mutant forms of oxytocin (Lee et al., 2011; Ren et al., 2015). The most common variant has a proline (non-polar, neutral) substituting for leucine in position 8 (Pro^8 oxytocin). The mutation is due to a single nucleotide change from thymine to cytosine (i.e., CTG to CCG). This change is found in marmosets, tamarins, owl monkeys, squirrel monkeys, spider monkeys, and capuchins, but not in titi monkeys and wooly spider monkeys, which have oxytocin with the consensus leucine at position 8 (Lee et al., 2011). This mutation likely has an effect on peptide structure. There is a correspondingly greater divergence in oxytocin receptor from the human form in New World monkeys expressing Pro^8 oxytocin (Ren et al., 2015). However, administration of the conserved form of oxytocin (Leu^8) to marmosets has the expected effects both centrally (e.g., Parker et al., 2005) and in stimulating milk let-down (M.L. Power, personal observation). In contrast, however, central administration of Leu^8 oxytocin failed to influence sociosexual behavior of marmosets toward unfamiliar opposite-sex conspecifics, while Pro^8 oxytocin reduced interactions with an opposite-sex stranger (Cavanaugh et al., 2014).

There are three other oxytocin variants in New World monkeys; two with variants at position 8 (alanine and threonine), and howler monkeys have a Leu^8 oxytocin with phenylalanine replacing tyrosine at position 2 (Ren et al., 2015). The evolutionary history of these mutations of the oxytocin gene is not completely transparent. All members of the Cebidae have the Pro^8 form of oxytocin; but so does the Atelid, the spider monkey. The other Atelids have the consensus Leucine in position 8, but howler monkeys have the change at position 2. The pithicids have either Leu^8, Ala^8, or Thr^8 oxytocin (Ren et al., 2015). New World monkeys show the greatest diversity in oxytocin sequence of all other mammalian groups so far studied.

Functions of Oxytocin

As befits an ancient hormone, oxytocin has been incorporated into many functions in different tissues. In addition to milk let-down and uterine contractions, peripheral oxytocin has mild vasopressin-like function on kidneys, a role in cardiovascular function, as mentioned above, and appears to inhibit inflammation and improve wound healing (Macciò et al., 2010; Poutahidis et al., 2013). The oxytocin receptor is expressed in skeletal muscle, and oxytocin appears to play a role in muscle growth and regeneration. Reduced circulating oxytocin in the elderly may underlie aspects of age-associated muscle degeneration (Kim et al., 2015). But its more studied functions are in the brain. Oxytocin is linked to social bonding, trust, and parental care.

Love Hormone?

Oxytocin has gained some recent notoriety in the popular press as the "love hormone," the "trust hormone," and even the "monogamy hormone." There is an entire industry producing oxytocin products to ingest or spray on clothes and skin or up your nose that purport to do anything from increasing the likelihood that people you interact with will trust you, to enhancing your own mood and sense of well-being. The product claims are based on the scientific understanding of some of the functions of oxytocin in the mammalian brain. Central oxytocin (oxytocin in the brain) is indeed associated with affiliative behaviors and trust, and it plays an important role in attachment. But the claims in the ads are more akin to using the language of science to support magical beliefs. Oxytocin didn't evolve as a pheromone, to be released into the environment to affect conspecifics. It is endogenously produced to act on the producer, not on others. Spraying it on your clothes is not likely to have much effect on the behavior of those around you.

On the other hand, infusions of oxytocin into the shell of the nucleus accumbens facilitates social contact and partner preferences in female prairie dogs through dopaminergic mechanisms (Liu and Wang, 2003). In male humans, a dose of nasal oxytocin increases the perceived attractiveness of a man's romantic partner; but not the attractiveness of strangers or acquaintances (Scheele et al., 2013). Many contexts in which social stimuli are evaluated activate central oxytocin expression (Kramer et al, 2006; Choleris et al, 2003; 2006; 2008). Experiments on lactating rats using functional magnetic resonance imaging (fMRI) demonstrate that oxytocin is associated with the mother-pup social bonding during suckling (e.g., Febo et al., 2005).

Oxytocin dampens amygdala activity; amygdala is involved in fear and antagonistic responses. Reducing fearful reactions is important in developing trust (Dreu, 2012). Peripheral oxytocin also may reduce anxiety. The injection of oxytocin subcutaneously in rats reduced background anxiety levels (Missig et al., 2010). Interestingly, intracerebroventricular injection of oxytocin did not reduce background anxiety (Ayers et al., 2011), suggesting that it was effects of oxytocin on peripheral physiology that reduced anxiety.

Intranasal administration of oxytocin has shown some positive effects on social attention and brain function in autistic children (Stavropoulos and Carver, 2013; Bakermans-Kranenburg and van IJzendoorn, 2013). An fMRI study of 17 children with high-functioning autistic spectrum disorder found that intranasal oxytocin enhanced activity in brain regions associated with reward and social attention and perception in response to pictures of eyes (social stimuli) but reduced activity in response to pictures of automobiles (nonsocial stimuli) (Gordon et al., 2013). However, modest at best improvements have been reported for social behavior and functioning in autistic children and adults given repeated doses of intranasal oxytocin (Anagnostou et al., 2012; Dadds et al., 2014).

Findings from Oxytocin Knockout Mice

Mice lacking the oxytocin gene can survive, but they display a mosaic of normal and abnormal behaviors and physiological responses. Sexual behavior does not appear to be affected. However, oxytocin knockout mice display significant social amnesia, failing to recognize other mice even after repeated social encounters (Winslow and Insel, 2002). They are also more aggressive toward conspecifics as adults.

Oxytocin knockout mouse dams will get pregnant and produce live litters. One reason an oxytocin-null strain of mice was created was to study the role of oxytocin in the timing and physical process of parturition. The thought was that pregnant mice without oxytocin would not go into labor; at least not efficiently. Surprisingly, oxytocin knockout pregnant mice gave birth just fine, with perhaps more variation in gestation length but no difference in mean gestation time (Young et al., 1996; Russell and Leng, 1998). The mothers appeared to exhibit reasonably competent maternal behavior at birth as well, in terms of licking and cuddling their pups. Thus, attachment behavior was not completely compromised. However, all the pups died. Why? Because the dams had no milk let-down reflex. The dams produced milk, but they could not express it; and the pups all starved to death. If the pups were fostered onto a lactating dam they would survive. Thus, for many functions of oxytocin compensation by other pathways appears to be possible; but not for milk let-down, at least in the mouse.

One of the main lessons from mouse knockout models in general has been an understanding of how often redundancy of function exists. Knock out a critical gene for a biological process and sometimes other pathways operate to produce, if not an identical biological result, at least a result comparable to the wild type. The apparent lack of any alternative pathways for milk let-down is intriguing, given the existence of an alternative mechanism for uterine contractions at parturition. Both events are critical for reproductive success in mammals. Contractions related to parturition is an older phenomenon; despite the evidence that the mammary gland and lactation are ancient, parturition and egg-laying, as well as parental behavior, are far older. It appears that 250–300 million years of evolution have not been sufficient to produce alternative pathways for milk let-down that do not require oxytocin, if the mouse result is true for all mammals.

Prolactin

In the late 1920s and early 1930s it was determined that a factor produced by the pituitary gland was necessary to stimulate milk secretion in multiple mammalian species (e.g., Stricker and Grueter, 1928; Corner, 1930; Nelson and Pfiffner, 1931). Riddle and Braucher (1931) demonstrated that a pituitary factor was responsible for the enlargement of the crop glands of pigeons and doves near the time of hatching and the subsequent production of crop milk. In 1933 it was shown that the same pituitary hormone was responsible for both milk secretion and crop enlargement, and it was given the name prolactin (Riddle et al., 1933).

Prolactin (PRL) has important physiological effects in support of mammalian lactation, a fact reflected in its name, which derives from this lactogenic function. But the prolactin gene (*PRL*) is ancient and predates lactation and even the existence of the mammalian lineage. Prolactin serves as an instructive example of how evolution often works, co-opting existing structures and signaling systems to achieve novel function. *PRL* is evolutionarily related to the growth hormone gene (*GH*); the genes for these hormones are the result of an ancient gene duplication of a single common-ancestor gene that existed in the vertebrate ancestor more than 345 million years ago. *PRL* and *GH* are found in all vertebrate taxa and perform a wide variety of often overlapping functions depending upon the taxa. Prolactin is an incredibly versatile hormone with hundreds of putative functions ranging from roles in fluid balance and osmotic regulation in fish to possibly a role in the regulation of metamorphosis in amphibians. Indeed, Dr. Carl Nicoll, an important researcher of prolactin function between 1959 and the 1990s, proposed changing the name to versatillin to reflect the diverse range of PRL function.

There have been more recent duplications within the PRL/GH gene family. In fish there are at least four surviving PRL/GH genes: *PRL1*, *PRL2*, *GH*, and somatolactin (*SL*). Of these four genes, mammals appear to have lost *PRL2*, and the advent of *SL* appears to have occurred after the fish lineage split from the tetrapod lineage. Thus, most mammal species appear to have only genes for prolactin and growth hormone. However, in rodents, ruminants, and primates there are entire families of polypeptide hormones with lactogenic and somatogenic properties, all derived from *PRL* or *GH*. Many of these variants are largely or even solely expressed in the placenta; the placental growth-hormone variants and the placental lactogens. These placental lactogens may exert important influences on the development of the mammary glands before birth.

Prolactin and Fluid Balance

The original functions of prolactin and growth hormone likely included osmoregulation; certainly in fish prolactin acts to regulate fluid balance. Many fish are faced with the challenge of adapting to both seawater and freshwater, either during different parts of their life cycle or, for some, on a shorter time scale, due to living in tidal areas at the mouths of rivers, where the salinity of the water can change dramatically because of tides or changes in river flow. Both prolactin and growth hormone have important functions in the necessary changes fish must undergo when they move between seawater and freshwater. In going from salty water to freshwater, fish use prolactin and growth hormone to regulate the reorganization of morphology and physiology to change the gills from an organ that excretes excess ions to one that absorbs these ions from the new, low-salinity environment.

In placental mammals, prolactin is linked with fluid balance for the amniotic sac. Hyperprolactemia is associated with polyhydramnios, and decidual prolactin may be one of many molecules that regulates permeability of the amniotic sac. Water deprivation results in an increase in circulating prolactin. Although prolactin does not appear to directly affect thirst and water intake, it does appear to act synergistically with angiotensin to increase water intake (Kaufman and Mackay, 1983). Prolactin also has function in regulating calcium absorption from the gut and excretion from the kidney (Wongdee and Charoenphandhu, 2012). The hyperprolactemia of pregnancy and lactation increases duodenal calcium absorption and decreases renal excretion in rats (Charoenphandhu et al., 2010, Wongdee and Charoenphandhu, 2012). Prolactin may act as a major calciotropic hormone during lactation (Charoenphandhu et al., 2010). Lactation has significant fluid and mineral balance consequences for the lactating female, and prolactin may act to regulate fluid and mineral balance in lactating mammals.

Evolution of PRL in Placental Mammals

The origin of the *PRL* gene predates mammals; indeed, it predates the origin of amniotes. The *PRL* gene in most mammalian orders exhibits strong conservation, probably indicating purifying selection. *PRL* exhibits little change from the reconstructed ancestral placental mammal gene for many lineages (e.g., dog = 4 amino acids, pig = 5 amino acids, horse = 10 amino acids, slow loris = 6 amino acids) and even in non-placental mammals such as the opossum there is little difference (19 amino acids). However, in at least four lineages *PRL* exhibits a mixture of periods of stasis with episodic bursts of rapid change, as does the prolactin receptor. Ruminant, rodent, and anthropoid primate (but not prosimian primate) lineages all display high rates of evolution for both *PRL* and its receptor. Interestingly, these are the same lineages known to have evolved placental lactogens, though these placenta-expressed genes derive from *PRL* only in ruminants and mice/rats. Placental lactogens in the anthropoid primates derive from *GH*, an interesting case of parallel evolution in which duplications of different genes that themselves came from an ancient gene duplication have resulted in families of peptides with convergent function. In anthropoid primates growth hormone is capable of stimulating the prolactin receptor; this is not true for ruminants and rodents. The gene duplications leading to placental-expressed PRL-like genes in these taxa occurred after the burst of *PRL* gene evolution in all these taxa.

In rodents, the placental lactogens have functions in maintaining the corpus luteum, fetal development, mammary development, and regulation of maternal energy metabolism. In humans, the placental lactogens appear to function to regulate maternal energy metabolism and do not have mammary development activity. The roles of the placental lactogens in ruminants are not fully understood, though they have lactogenic activity (Takahashi et al., 2013).

The elephant and hyrax lineages of the mammalian superorder Afrotheria (but not the tenrec lineage) also show high rates of evolution of *PRL*,

---

BOX 4.1

Synonymous versus nonsynonymous mutations

A mutation in the DNA sequence results in a change in one or more nucleotides. If that change occurs within a codon (three adjacent nucleotides that code for a particular amino acid) of an expressed (transcribed and translated) portion of the DNA, it may or may not result in a difference in the expressed gene product. The four base nucleotides (guanine [G], adenine [A], thymine [T], and cytosine [C]) can combine to form 64 possible codons. Three codons (TAG, TAA, and TGA) are called stop codons. They do not code for an amino acid but rather bind to release factors that separate the amino acid chain being built from the ribosomal subunits, thus terminating translation. There are 20 amino acids that are coded by the remaining 61 codons, with as few as a single codon coding for tryptophan (TGG) to as many as 6 codons that code for the amino acid leucine. A synonymous mutation is a change in a codon that does not result in a change in the amino acid; a nonsynonymous change means that the amino acid being coded for by the mutant codon differs. For example, CTT and CTC both code for leucine, however CCC codes for proline. A mutation of CTC to CTT would be a synonymous mutation while a mutation from CTC to CCC would be a nonsynonymous mutation.

Synonymous mutations were originally considered silent and would have no selective advantage or disadvantage. This has been shown not to be true in all circumstances. Some codons are translated faster and/or more accurately than others. Different codons that code for the same amino acid can still affect the *(continued)*

---

but it doesn't appear that any *PRL* gene duplications occurred, or at least none that were retained. Within Afrotheria, the tenrec *PRL* shows little divergence from the reconstructed ancestral placental mammal *PRL*, only differing by 8 amino acids, while elephant *PRL* differs at 49 amino acids and hyrax at 94 (47%) amino acids (Wallis, 2009). Evidence suggests that the rapid evolution of *PRL* in hyrax and elephant occurred due to adaptive selection. What drove the rapid bursts of evolution and why these bursts were not followed by gene duplication events to produce placental lactogens is unknown; but all of the functions of prolactin are not well understood for these species.

There are few nonsynonymous coding sequence (box 4.1) changes in lineages with conserved *PRL* sequences, such as opossum, tenrec, rabbit, pig, dog, and the prosimian primate the slow loris. In the lineages that display

BOX 4.1 CON'T

rate at which the final peptide is produced, potentially affecting phenotype. For example, fruit flies with synonymous mutations of the alcohol dehydrogenase (*adh*) gene showed wide variation in ADH activity, presumably due to different rates of ADH production (Carlini and Stephan, 2003). Synonymous mutations can affect the half-life of the messenger RNA (mRNA) as well as the folding of the mRNA and hence the rate of translation. Both can affect the rate of peptide production. In addition, the rate of translation can affect the folding of the protein being produced, which can have profound effects on its function. Synonymous mutations, predicted to be neutral by population genetics theory, may have selective consequences.

Nonetheless, nonsynonymous mutations have a greater probability of affecting phenotype and being subject to natural selection. Many synonymous mutations truly are silent. The ratio of nonsynonymous-to-synonymous mutations can indicate whether purifying selection is acting (strong pressure to retain the composition of the peptide, which means the ratio will be very small), or selection is driving adaptive change in function or structure of the peptide, in which case the ratio may be greater than 1. If the peptide has lost function, and thus is not under selection, the prediction is that the rate of non-synonymous and synonymous mutations will be approximately equal. Figure 4.1 illustrates a relatively constant rate of synonymous nucleotide substitutions in the PRL gene among mammals; but much greater nonsynonymous rates among rodents, ruminants, anthropoid primates, and the elephant and hyrax, indicating probably bursts of selective change in PRL in those lineages.

episodic rapid evolution of *PRL*, such as the lineages leading to elephants, rats and mice, ruminants, and anthropoid primates, there are many more nonsynonymous sequence changes (figure 4.1).

Functions of Prolactin in Anthropoid Primates

The rapid evolution of *PRL* in the anthropoid primate lineage appears to have occurred early on, with the time period being after the split from the prosimian primates but before the divergence of the New World monkeys. The molecular evidence suggests that adaptive change in the *PRL* sequence occurred before the duplication of the *GH* gene that led to the placental lactogens in anthropoid primates. This implies that prolactin may have taken on a new function during this time. The duplication of *GH* following this adaptive evolution of *PRL* allowed the divergence of anthropoid *GH* from

(a) Synonymous   (b) Nonsynonymous

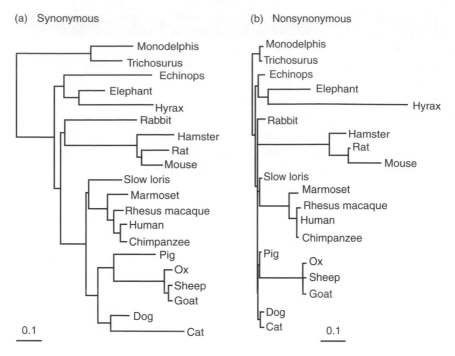

Figure 4.1. Rapid PRL gene evolution; synonymous and nonsynonymous codon changes. From Wallis, 2009; reproduced with permission.

the generally conserved ancestral sequence and the evolution of new functions by some of the duplicated genes, including some lactogenic function.

What might that new function for prolactin have been? One possibility is a role in decidual formation and maternal immune tolerance to the implanting blastocyst. Prolactin certainly plays an important role in these processes during human pregnancy. Prolactin acts on uterine epithelium to facilitate decidua formation in multiple ways. It is involved in uterine epithelial cell differentiation, promotes trophoblast growth from the implanting placenta, and promotes maternal uterine angiogenesis at the implantation site. Prolactin down regulates both IL-6, an inflammatory molecule, and 20α-HSD, which catabolizes progesterone. Thus, prolactin regulates inflammation and increases progesterone in uterine tissue. Progesterone is required for decidua formation and successful implantation. Prolactin also modulates the maternal immune response, playing a role in the switching of uterine natural killer cells from phagocytic to regulatory in nature. Thus, the implanting blastocyst is not attacked by the maternal immune system, but rather finds a hospitable uterine epithelium that actively signals and responds to signals from the blastocyst, coordinating the adhesion and then penetration of the epithelium for interstitial implantation (Lash et al., 2010; Power and Schulkin, 2012).

Expression of *PRL* in endometrium is not found in all other mammalian lineages; rabbits, pigs, dogs, and armadillos do not express *PRL* in uterine tissue. Nor, however, is it unique to anthropoid primates, as endometrial *PRL* expression occurs in mice and elephants. Thus it occurs in phylogenetically dissimilar taxa, with elephants being members of the superorder Afrotheria, which diverged from the lineage leading to primates close to the origin of placental mammals. The promoters for uterine expression of *PRL* in these taxa are different from the pituitary *PRL* promoter, but also quite different from each other. In humans the uterine *PRL* promoter is derived from the long terminal repeat (LTR) element *MER39*. Since this is also true for the spider monkey, a New World monkey, it appears that there was a single event that occurred prior to 40 million years ago that resulted in this alternative promoter for *PRL* that regulates uterine expression. In contrast, the mouse promoter derives from LTR element *MER77*, and the elephant *PRL* promoter from the lineage-specific LINE retrotransposon *L1-2 LA*, implying that these adaptive events were independent, convergent evolution (Emera et al., 2012). Interestingly, all three of these lineages (anthropoid primates, murid rodents, and elephants) display a period of rapid *PRL* evolution in their history. The hypothetical new function for prolactin that might underlie such rapid evolution may indeed have been related to uterine epithelial expression and implantation (Wallis, 2009). Uterine *PRL* expression in the fourth lineage to have episodic rapid *PRL* evolution, the ruminants, has not yet been examined.

Extrapituitary expression of *PRL* is common in primates. Besides decidual expression, *PRL* is also expressed in skin, hair follicles, kidney, testes, prostate, thymus, and spleen. Primates have an additional *PRL* promotor upstream of the pituitary transcription site (Berwaer et al., 1994). Mice bioengineered to express human DNA containing the human *PRL* gene and promotor sites as well as the mouse *PRL* gene expressed human *PRL* in many extrapituitary sites, but expressed mouse *PRL* only in the pituitary (Christensen et al., 2013).

Extrapituitary Expression of Prolactin

Although the pituitary is the main source of circulating prolactin, prolactin is produced by other tissue and has important autocrine functions. *PRL* is expressed in human decidua, ovaries, immune system glands (thymus, spleen, tonsils, and lymph nodes), hair follicles, and adipose tissue. *PRL* is expressed in male rodent reproductive organs (testes and prostate), the cochlea, and hair follicles. The *PRL* gene is expressed in vascular endothelial cells; however, the product is subject to post-translational processing that produces biologically active fragments of prolactin which have anti-angiogenic activity (Horseman and Gregerson, 2014; Marano and Ben-Jonathan, 2014). The activity of these fragments is not transduced through the prolactin receptor (Clapp and Weiner, 1992).

PRL is expressed at low levels by human mammary gland epithelial cells. Breast adipose tissue expresses *PRL* at higher levels than does breast glandular tissue (Zinger et al., 2003). The origin of prolactin from adipose is not fully understood; but adipose tissue macrophages certainly express *PRL* (Bouckenooghe et al., 2014). Prolactin expression by adipose tissue macrophages is upregulated by both glucose and by inflammatory signals (e.g., IL-1β), with the effects being additive, suggesting a possible role for prolactin in the development of diabesity (Bouckenooghe et al., 2014).

In the mouse, autocrine action by mammary prolactin is required for successful lactation. Prolactin expression in the mammary gland depends on activation of the PI3K-Akt1 pathway. The subsequent phosphorylation of STAT5 via the PRL-JAK2 pathway results in terminal mammary epithelial differentiation and milk production. Blocking the synthesis of prolactin by the mouse mammary resulted in lactation failure, despite normal pituitary prolactin release (Chen et al., 2012).

Extrapituitary expression of *PRL* appears more common in mammals than in the sauropsid lineage. Modern sauropsids, such as lizards and birds, appear to express prolactin only in pituitary, brain, and reproductive organs (ovaries and testes), although the prolactin receptor is found in most tissues (Kato et al., 2005). The production of prolactin by mammalian skin is intriguing and relevant to the evolution of the mammary gland and lactation and of hair. Mammary glands can be considered an integumentary appendage, just like scales or feathers in sauropsids and hair in mammals. Prolactin does not appear to be significantly involved in skin in squamate reptiles and birds; it is important in both hair and mammary glands. An important role for prolactin in development and diversification of the integument for synapsids may have been a critical preadaptation for developing mammary glands (and hair), and may represent an important adaptive divergence between the ancestors of mammals and the extant reptiles.

Parental Care

Prolactin is important in many aspects of postpartum reproductive behavior in a wide variety of vertebrates (Horseman and Gregerson, 2014). Prolactin is required for crop milk production in pigeons and doves and for skin mucus production in discus fish (see chapter 2). Thus, prolactin is involved in secretions fed to offspring in animals phylogenetically quite distinct from mammals.

Prolactin is a molecule associated with parental behavior in a number of taxa. It is linked with brooding behavior in birds and both maternal and paternal behavior in mammals. Physical contact is associated with circulating prolactin levels in several mammals. Physical contact with infants increases circulating prolactin in male marmoset monkeys (Dixson and George, 1982). Male marmosets provide substantial infant care, carrying infants and even provisioning them with food. Even inexperienced male marmosets will retrieve unfamiliar infants; retrieving infants was associated with an increase in urinary prolactin levels (Roberts et al., 2001).

Prolactin is also linked to transitional states between daytime and dark periods; like melatonin, it is knotted to dark periods and under circadian control (Brainard et al., 1981; Wehr et al., 1993). Prolactin infusion into the central region of the amygdala facilitates REM sleep (Sanford et al., 1998). Prolactin is a calming hormone, leading to quiescence in animals. This action of PRL is conducive to a female suckling her young by calming her, reducing the motivation for activity and aggression.

Prolactin and Metamorphosis

Prolactin has somatotrophic and developmental functions in many species. Its role in metamorphosis in amphibians is uncertain, as its actions may derive from its similarity to growth hormone and its ability to stimulate the growth hormone receptor. However, prolactin is linked to the reorganization and phenotypic change of organs. It functions to reprogram fish gill tissue when fish move between fresh and salt water, reprograms uterine tissue in many placental mammals, and, of course, it acts on mammary tissue in female mammals before and after birth to transform a quiescent gland into an actively secreting gland. Because of its somatotrophic effects and its regulatory effects on expression of other molecules related to inflammation and immune function, prolactin can have multiple phenotypic effects depending on the organ system it is acting upon. The metamorphic properties of prolactin in essence preadapted it to function in changing structure and function of the mammary gland.

Prolactin can affect tissues in a variety of ways, both directly and indirectly. A potentially powerful mechanism is through its effects on stem cells. Stem cells are found in most if not all organs of adult mammals. These stem cells, termed adult stem cells or somatic stem cells to differentiate them from embryonic stem cells (box 4.2), are important for repair of tissue damage in many organs, and the normal renewal of tissue for such organs as the intestines. A transplanted mouse mammary stem cell was able to grow into a functional mammary gland (Shackelton et al., 2006). Prolactin has been shown to regulate adult stem cells in a variety of tissues, both *in vitro* and *in vivo*. Prolactin has tissue-specific effects on stem cells, maintaining quiescence in some and promoting proliferation and/or differentiation in others (Sackmann-Sala et al., 2015). Adult stem cells in the mammary gland are essential for the transformation from a quiescent gland to a milk-secreting gland.

Prolactin and Milk Production

Prolactin induces lobuloalveolar growth in the mammary gland and stimulates lactogenesis (Hennighausen et al., 1997). It stimulates *WAP* expression and, in combination with insulin and glucocorticoids, β-casein production in mouse mammary epithelial cells (Bolander et al., 1981; Hennighausen et al., 1997). Prolactin acts by binding to the prolactin receptor and initiating signaling cascades. Signal transduction is mainly through the Jak2-Stat5 pathway (Chen et al., 2012), though PRL can also activate Stat1 and Stat3 (Hennighausen et al., 1997) and also signals via the MAPK pathway. After phosphorylation by Jak2, Stat5 translocates to the nucleus of mammary epithelial cells to bind to response elements and regulate gene expression (Hennighausen et al., 1997; Chen et al., 2012). Stat5 signaling is required for the mammary gland to develop appropriately during pregnancy, at least in the mouse (Hennighausen et al., 1997). It also regulates the expression of many, but not all, milk proteins. Loss of Stat5 activation results in lactation failure.

There is variation in the *PRL* gene in bovines. Cows have at least two *PRL* alleles (A and B), and milk production varies according to these alleles. Cows of type AA produce the most milk, followed by AB cows (Alfonso et al., 2012; Ghasemi et al., 2009). Administration of recombinant human

BOX 4.2

Stem cells

Stem cells are undifferentiated cells that can retain their undifferentiated state after mitosis. When a stem cell undergoes mitosis it typically produces one daughter cell that becomes differentiated, and acquires specific function/phenotype, and a second daughter cell that remains undifferentiated. However, the other two possibilities (two differentiated daughter cells or two undifferentiated daughter cells) also occur.

Embryonic stem cells typically come from the inner cell mass of blastocysts while adult stem cells (also known as somatic stem cells) come from tissue from organisms after birth. Beyond their separate origins, embryonic and adult stem cells tend to have different properties, especially regarding what is termed potency. Embryonic stem cells are descendants of what are termed totipotent (or omnipotent) stem cells, which are capable of differentiating into any cell type. Embryonic stem cells are typically pluripotent, which means they can differentiate into most but not all cell types. Adult stem cells tend to be multipotent, which means they can differentiate into many cell types, but usually within a set of related cell types. Some adult stem cells are unipotent, which means they can continue to self-replicate and produce undifferentiated cells, but when they do differentiate they produce only one cell type. Adult stem cells are important for tissue repair and regeneration of organ tissue. The adult stem cells found in an organ can differentiate into the cell types found in that organ or produced by that organ. For example, bone marrow stem cells can differentiate into any of the blood cell types.

Adult stem cells are important for mammary gland development and function. It has been shown that a single mammary stem cell can generate a complete, functional mammary gland (Shackelton et al., 2006). However, mammary glands also contain unipotent stem cells in addition to a lineage of multipotent stem cells (Visvader and Stingl, 2014), so mammary stem cells are a heterogeneous mix. Beyond any function in tissue repair, which is important for any organ or tissue in the body, mammary stem cells provide the potential to become a secretory organ. The mammary gland cycles between being a quiescent organ and an actively secreting organ. During its secreting state it contains a population of secretory cells that are lost to apoptosis at the end of lactation (involution). The population of mammary stem cells allows the gland to retain the ability to become once again a secretory organ after the next pregnancy.

prolactin to mothers with prolactin deficiency or to mothers of preterm infants with lactation insufficiency increased both milk volume and lactose concentration of milk (Powe et al., 2011). An increase in lactose synthesis would be expected to increase milk volume due to its osmotic properties (see chapter 6 for a more complete discussion).

Why Don't Male Mammals Lactate?

Lactation may be a defining characteristic of mammals; but it appears to be solely a female trait. Male mammals do not, under normal circumstances, lactate. Males of many, but not all, mammalian species have the anatomical machinery. Male rodents lack nipples after birth. Nipples are formed in utero, but embryonic release of testosterone leads to a cascade event resulting in the nipples degenerating and being lost. Stallions also lack nipples. In primates, mammary development does not differ between males and females until puberty, so men have the basic anatomical structures for functional (lactating) mammary glands. What men generally lack is the appropriate hormonal milieu. Men with pituitary tumors that result in prolactemia have produced small quantities of a milky discharge from their nipples (e.g., Anoop et al., 2010). But almost all examples of male lactation have been unusual circumstances of drug use, disease (especially tumors), or extreme deprivation. For example, many male survivors of Nazi and Japanese prison camps experienced a milky discharge from their nipples after they were rescued (Swaminathan, 2007). Due to their extreme malnourished state their glands and liver had largely shut down. After they were rescued and returned to proper nutrition, their glands recovered more quickly than their livers. The liver is important for the metabolic breakdown of hormones, including prolactin, and the imbalance between returning liver function and gland function caused the men to become hyperprolactemic (Diamond, 1995).

The only male mammals that have been observed to lactate under natural circumstances are two species of fruit bats. Male Dayak fruit bats appear to lactate. They produce a milky discharge from their nipples; but milk production is much smaller than for female bats. Nipples of males do not show the changes associated with nursing that are seen in nursing female Dayak bats (Kunz and Hosken, 2009). It is not clear if any male milk being produced is actually ingested by pups. Many of the plant foods eaten by these fruit bats contain phytoestrogens, so male lactation may again merely represent a hormonal dysregulation.

We are not aware of any cases in which the milky discharge produced by males has been shown to have the same composition as true milk of the species. The male mammary gland of some species can be induced to become a secretory gland; but it isn't clear that the secretion will contain the entire biochemical complexity of milk produced by a lactating female. In all cases the volume of secretion produced is much lower than is typical for lactating females.

Perhaps the answer to why males don't lactate lies in the ancient origins of lactation. If Oftedal is largely correct, lactation first served as an egg-protecting behavior. Perhaps the stimulation of proto-milk secretion was tied to oxytocin and oviduct contractions. Since male mammals never give birth (or lay eggs) this connection does not exist. A male mammal can be caused to lactate, by external hormonal treatments or tumor-driven prolactemia, but true lactation remains a female trait.

Oxytocin and Prolactin Together

Both of these ancient hormones have been co-opted to serve lactation. Their genesis and ancient functions preadapted them to be able to serve this novel adaptation in the synapsid lineage. Oxytocin is a molecule of parturition and prolactin a molecule of postpartum parental care. They are also linked by regulatory mechanisms. Oxytocin is a stimulator of prolactin release, while dopamine is inhibitory to prolactin release by pituitary lactotrophs (Kennett and McKee, 2012). Prolactin stimulates oxytocin secretion, initiating a positive feedback loop; however, prolactin secretion has also been shown to inhibit oxytocin neurons (Kokay et al., 2006), suggesting it can inhibit its own stimulatory mechanism directly (Kennett and McKee, 2012). The interlocking loops between oxytocin-prolactin (positive feedback) and prolactin-dopamine (negative feedback) reflect the complexity of regulatory systems in which knowledge of any one loop may not be especially instructive or predictive of the physiological responses. It is the wonderful, evolved complexity of interlocking and overlapping stimulatory and inhibitory feedback loops that create complex physiology and behavior.

Mivart Finale

Mivart's objections to the evolution of lactation were logically based on the notion that all evolution is slow and gradual. The initial secretion from which milk derives could not have had any particular function or complexity in his opinion. Science has showed such reasoning to be flawed. The concept of slow, gradual change being the main process by which evolution works has been challenged repeatedly (e.g., Eldredge and Gould, 1972; Gould and Eldredge, 1977; Newman et al., 1985). The fossil record exhibits many instances of periods of apparent morphological stability followed by bursts of rapid change. Adaptations often evolve through gains of function rather than true *de novo* evolution of novel ability. The precursors necessary to gain new function were already in place, allowing rapid change.

Many aspects of lactation are extremely ancient in origin; they were co-opted to serve new functions in lactation, but did not have to be derived *de novo*. Oxytocin already served to stimulate contractile function related to birth (or egg laying) and was involved in stimulating maternal behavior. Prolactin was also involved in maternal behavior, as well as possible regulation of the function and structure of tissues, including skin. The skin already had changed in amniotes to be able to secrete lipid. The early synapsids used glandular skin to combat water loss, as opposed to the sauropsids that evolved novel skin proteins (β-keratins). The components to support cutaneous glandular secretions tied to reproduction were in place. If we could travel back in time and sample proto milk from an early synapsid and examine the hormonal and metabolic signals involved in the regulation of proto lactation, it would not be surprising to find a fairly complex system from the beginning. The proto milk might not look like any modern milk; but we would likely recognize many of the precursors of modern milk constituents. Regardless of how nutritious the secretion might have been, proto milk and proto lactation undoubtedly were complex adaptations that served important adaptive functions, both for the mother, the eggs, and possibly hatchlings. Over the next hundred million years lactation and milk evolved to become similar to the modern milk of monotreme, marsupial, and placental mammals, and continued to evolve to arrive at the diversity and complexity of lactation and milk of modern mammals.

PART II

Milk as a Food

The primary adaptive purpose of milk is to transfer resources from a mother to her offspring so that they can grow and develop into independently feeding individuals. The transferred maternal resources traditionally have been considered to be nutritional. Despite our emphasis in this book on the signaling and regulatory aspects of milk, a fundamental function of lactation is to deliver nutrients to a female's offspring. Milk has to be a food; it is just not only a food. In this section we restrict our examination of milk to its nutritional function.

Milk provides the necessary nutritional components for metabolism, growth, and development for a mammalian neonate for at least a certain period of time. This is especially true for the monotreme and marsupial species, which are born extremely small and undeveloped, and grow and develop for an extended time sustained only by milk. Among placental mammals, most neonates have significant stores of some nutrients that were passed from their mothers through the placenta. For example, neonates are generally born with substantial stores of iron and vitamin D (discussed further in chapter 5). But milk provides the entire diet of all mammalian neonates for some period of time. Milks of all species must provide the following: a source of metabolizable energy (e.g., fat and sugar), the basic building blocks for growth (e.g., protein, calcium, and phosphorus), micronutrients necessary for metabolism (e.g., minerals such as zinc and sodium, as well as vitamins), and water. Thus, it isn't surprising that all milks have the same basic nutrients in common: water, lipids (fat), proteins, sugars (often mostly lactose), minerals, and vitamins. The proportion of water, fat, sugar, protein, and minerals, however, varies widely among species (Langer, 2008; Oftedal and Iverson, 1995). This has implications for understanding the reproductive strategies of species, the differences in patterns of growth of neonates, as well as practical implications for managers of captive animals that are often faced with the necessity of hand-rearing a wide variety of mammalian species.

In chapters 5 through 7 we focus on the constituents in milk that support the energetic and nutrient costs of metabolism and growth; milk as a food or, more properly, milk as a complete diet. The purpose of the chapters in this section is to give the reader a good understanding of the extent of variation in nutrient composition of milks both between different species as well as within species in terms of variation across lactation and between individuals, as well as the types of constraints on milk composition. Not all combinations of fat, protein, and sugar can be found in milk. Some are biochemically impossible (or at least extremely unlikely). Others would not be able to support mammalian life and growth. Milk composition may vary widely, but there are patterns and apparent constraints.

The Perfect Food?

Common perception is that breast milk is the perfect food for babies. But evolution and perfection rarely go together. Evolved solutions to challenges are inherently different from designed ones. Evolved adaptations are solutions to past challenges. Evolution does not predict future challenges. For each mammalian species, the nutrient composition of the milk represents a blend of phylogenetic constraints that represent the distant past with more recent evolutionary events in that species lineage. In addition, adaptations often represent compromises between competing challenges. Nutrition, though critically important, is not the only parameter vital for neonatal success. And milk is expensive for the mother to produce. The fitness imperatives of mothers and babies are not always perfectly aligned, so what might be perfect for the baby may not be perfect from the mother's fitness perspective. Mother's milk represents an evolved compromise between many selection pressures.

Chapter 5 will address the question: In what aspects does mother's milk appear not to be perfect as a food, and why? It will also consider the extent to which mothers may modify their milk so that it differs for different offspring, possibly even depending on the sex of her offspring.

The Milk Nutrient Spectrum

All neonates require energy and essential nutrients, such as minerals and vitamins. Some of these nutrients are metabolized (e.g., lactose and most lipids), and others are more likely to be deposited (e.g., calcium and phosphorus into bone). As a broad generalization, the proteins and minerals in milk can be considered the building blocks by which an infant grows, while the lipids and carbohydrates provide the energy to support growth processes. Like any generalization there are many exceptions. Some lipids (e.g., long-chain polyunsaturated fatty acids) are often incorporated into tissue and not metabolized for energy; some proteins may be metabolized and provide energy; and of course some peptides are signaling molecules or immunoglobulins, with functions in metabolism, physiology, development, and health outside of nutrition. Finally, while lactose in milk certainly is a source of metabolizable energy, we are just beginning to investigate the functions of other carbohydrates found in milk, the milk oligosaccharides.

Perhaps the simplest method of distinguishing between milks is whether the main energy source is sugar or fat. Indeed, across mammals fat content is the most variable aspect of milk composition. There are species with very low fat content (less than 1%; e.g., horses and rhinoceroses) and species with as much or more fat in their milk as there is water (e.g., many seals). The champion high-fat milk comes from the hooded seal, with a fat content of 54–60% (Oftedal et al., 1993a).

High-fat milks are low in sugar; high-sugar milks are low in fat. There are species that produce milk with moderate sugar and fat. But there are no high-fat, high-sugar milks. There are constraints on milk composition. Some constraints are biochemical, some are biological, and some may be phylogenetic. In chapter 6 we explore the variation in placental mammal milk composition, looking for patterns and possible explanations.

Finally, we examine the technical question of how best to compare milks that may differ greatly in apparent composition. We frame this exploration around comparing milk constituents between two species (Asian elephants and the white rhinoceros) that appear at first glance to have substantially different milks. But is elephant milk really that different from rhinoceros milk? The answer depends on how you express the values for how much of any constituent is in the milk. Using one set of units the milks will look different; but change the units of measurement and some of the differences can disappear. The important judgment for the researcher is: what units are the most likely to illuminate the important selective/adaptive features of the milks? How can we measure the composition of milk so that it best represents its biologically functional values?

Lactation Strategies

Milk composition derives from and has consequences for the life history strategy of a species. Chapter 7 will outline what comprises a lactation strategy, and how milk composition has evolved to match the wide variety of mammalian life history strategies.

The transfer of maternal resources to offspring through milk depends on four basic parameters: the composition of the milk, the volume of milk produced per day, the frequency of suckling by the offspring, and the total length of lactation. All of these parameters interact with the growth rate, developmental pattern, and digestive and metabolic capabilities of the offspring to form the species' lactation strategy. The diversity of existing mammals has resulted in a wide range of lactation strategies, and thus milks of widely varying compositions. The length of lactation is also highly variable across mammals. A hooded seal cow may produce a very high-fat milk, but she only does so for a short time. She nurses her pup for less than four days, and then leaves for good (Oftedal et al., 1993a). In contrast, elephant and great ape mothers will nurse their calves/infants for three to five years or more. The length of lactation is obviously strongly associated with the age at which offspring become independent, and also with the interbirth interval for the mother. A long lactation allows a lengthy, slow development of the offspring, but also reduces the number of births the mother can produce over her life.

Mammalian neonates can be highly altricial (poorly developed), for example in carnivore species where neonates are born essentially helpless, with eyes and ears closed. Or they can be highly precocial (well developed), such as in horses, giraffes, and many other hoofed mammals where neonates are capable of walking within hours of birth. Growth and development can be rapid, as in many rodent species, or rather slow, as in primates and especially humans. Obviously the lactation strategy of a species must be compatible with the pattern of growth and development of its young. But there are other factors, such as diet, predation risks, maternal care patterns, and so forth that influence lactation strategy. For mammals, the lactation strategy of a species is central to its life history strategy. In chapter 7 we examine the different lactation strategies that have evolved and explore the correlates with other aspects of life history strategy.

Not Quite Perfection

All mammals start life as lactivores, whether as adults they will be carnivores, herbivores or omnivores. Milk is not only their first food, it comprises their entire diet for some length of time. Milk has evolved to suit the growth and development of the offspring of each species. It is the best food for a neonate; or, at least, the best food that exists in nature. But there are still challenges to being a lactivore, and milk is not the perfect food under all circumstances. A species' milk must be understood within the evolutionary context under which it evolved. Milk is a well-adapted food for neonates, but it evolved as a balance between neonatal need, maternal capacity, and other factors with selective significance. Milk represents an evolved compromise between competing selective pressures, and sometimes those different selective pressures favor different concentrations of a nutrient or other bioactive factor in milk. The resulting milk composition evolved because, on balance, it achieved the highest reproductive success given the circumstances; but evolution doesn't predict novel selective pressures. New challenges faced by future neonates are not anticipated. If the environment is changed from the evolutionary past then milk that was sufficient then may be lacking now. The milk that was the best for overall fitness may no longer meet all requirements. If these new challenges favor a different milk composition, then in time, if the species survives, a change will likely occur; but only after many, many generations.

This last fact is especially relevant to aspects of human breast milk composition. Human breast milk is marginal if not deficient in iron and vitamin D, and both iron and vitamin D deficiency, unfortunately, are not uncommon in exclusively breastfed infants (Calvo et al., 1992; Merewood et al., 2012; Maguire et al., 2013). This would certainly seem to challenge the idea that breast milk is the perfect food. To understand why human breast milk contains low concentrations of these two important nutrients we turn to our evolutionary history. In one case (iron), the answer is likely due to competing selective pressures. There are advantages to low iron content in milk

BOX 5.1

Vitamin or hormone? A primer course on vitamin D

A vitamin, by definition, is a molecule necessary (in very small amounts) for metabolism that is not synthesized by the body and must be ingested. Thus, whether a substance is truly a vitamin depends not only on its structure and function, but also on the species of animal in question. For example, vitamin C is a vitamin for people, all other anthropoid primates (monkeys and apes), fruit bats, and guinea pigs. If individuals of these species do not ingest vitamin C they will develop scurvy. But most other mammals produce vitamin C in their liver, and so it is not required in their diets. It doesn't do them any harm if they ingest it, but they don't need to. For most mammals vitamin C is an endogenously produced enzyme, not a vitamin.

By this definition vitamin D is not a vitamin for humans, and for most other mammals that have been studied. Primates, rodents, artiodactyls (sheep, cows, and pigs) among other mammals all are capable of producing vitamin D in their skin in the presence of unfiltered sunlight or artificial light that contains the appropriate wavelengths of ultraviolet light (UV-B). There are carnivores that do require dietary vitamin D (e.g., cats; see box 5.2), and we don't really know how the many nocturnal or fossorial (underground-living) mammals obtain their vitamin D. For many of those species it may also be a vitamin; not necessarily because they can't produce it, but because their lifestyle greatly restricts the amount of sunlight they are exposed to so that they produce little or none. For most other mammals the evolutionary truth is that vitamin D is not a vitamin but rather it is an endogenously produced secosteroid precursor to a steroid hormone.

There are two forms of vitamin D in nature: cholecalciferol or vitamin D_3 which is produced by animals, and ergocalciferol or vitamin D_2 which is produced by fungi. Humans can metabolize either form into the active metabolite that has biological function (see below), although vitamin D_3 has been shown to have greater potency. Some species, such as some New World monkeys, can only utilize vitamin D_3 and will develop vitamin D deficiency if fed only vitamin D_2, (and not exposed to UV-B light).

Vitamin D_3 is made from a cholesterol precursor, 7-dehydrocholesterol. To be specific, it is made by a series of biochemical steps that starts when 7-dehydrocholesterol (also called provitamin D) reacts with light at a fairly narrow range of wavelengths in the UV-B spectrum (270–300 nm, with peak production at 295–297 nm) and is converted to previtamin D, an unstable molecule that is quickly converted to vitamin D_3 at normal skin temperatures. (In fungi, the sterol ergosterol is photosynthetically converted by UV-B radiation into vitamin D_2.) The same UV-B wavelengths that produce vitamin D also cause sunburn *(continued)*

BOX 5.1 CON'T

and potential damage to the DNA in cells. Indeed, Holick (2003) has suggested that the conversion process of 7-dehydrocholesterol (or ergosterol) to vitamin D may have originally evolved in single-celled organisms very early in Earth's history as a de facto sunscreen. Phytoplankton and zooplankton have the capacity to produce vitamin D in the presence of UV radiation. The vitamin D system probably extends back close to one billion years.

In humans, vitamin D_3 in the skin is quickly absorbed into the bloodstream and binds to plasma vitamin D-binding protein. Vitamin D_3 has very little biological activity. Two hydroxylation steps are required to turn vitamin D_3 into its biologically active form, 1,25 dihydroxyvitamin D (1,25$(OH)_2$D), also known as calcitriol. The first hydroxylation occurs in the liver, transforming vitamin D to 25 hydroxyvitamin D (25(OH)D). This metabolite is the longest lived in the body, and circulating levels in the blood are the best indicator of vitamin D status. The second hydroxylation occurs in the kidney, transforming 25(OH)D into 1,25$(OH)_2$D, a steroid hormone that binds with high affinity to and activates the vitamin D receptor (VDR).

Health risks of low vitamin D

Low vitamin D status is associated with increased risk for a number of diseases besides the bone diseases of rickets and osteoporosis. There is a strong association between low vitamin D status and colorectal cancer (Garland and Garland, 1980; Grant and Garland, 2004; Jenab et al., 2010). Indeed, low vitamin D status appears to be a risk factor for many cancers; not surprising considering that 1,25-dihydroxyvitamin D is a potent hormone in the regulation of cell growth (Zhang and Naughton, 2010). Low circulating 25-hydroxyvitamin D is associated with an increased risk of cardiovascular disease (Anderson et al., 2010). Vitamin D supplementation reduces inflammation, possibly by its effect on cytokine profiles, increasing anti-inflammatory cytokines such as IL-10 (Schleithoff et al., 2006). Although the mechanisms are uncertain, low vitamin D status is associated with poor glucose metabolism, impaired insulin secretion, and insulin resistance (Alvarez and Ashraf, 2010). Increasing circulating 25-hydroxyvitamin D levels improved insulin sensitivity in obese women (Tzotzas et al., 2010). Thus, vitamin D insufficiency appears to be linked with the development of diabetes, both type 1 and type 2 (Osei, 2010). Vitamin D appears to have potent effects on immune function, and low vitamin D status increases the risks of contracting infectious disease, including higher risk for contracting tuberculosis (Yamshchikov et al., 2009; Zhang and Naughton, 2010). Overall mortality risk declines with increasing circulating 25-hydroxyvitamin D up to a threshold of 75–87.5 nmol/L (Zittermann et al., 2012). Finally, low vitamin D status is associated with significant cognitive decline in the elderly (Llewellyn et al., 2010).

in terms of reducing the growth of pathogenic microbes in milk and in the infant's intestines. For vitamin D the answer is that for most of human evolutionary history our ancestors produced all the vitamin D they needed through a photosynthetic reaction in the skin when exposed to sufficient sunlight (box 5.1). Vitamin D is actually a fat-soluble secosteroid precursor to a steroid hormone, not a vitamin. Our ancestors didn't need it in their diet. Human circumstances have changed, and vitamin D has now become a necessary vitamin for many humans due to the nature of our modern environment, and our own choices on how we want to live.

We discuss these and a few other examples of ways in which milk does not appear to be "perfect" in more detail below. The important point is to illustrate that evolution has often favored compromise solutions that balance multiple selective pressures, and that were adaptive under past circumstances.

Vitamin D

Milk is often called the perfect food. However, milk is actually deficient in several nutrients. A prime example is vitamin D. Milks from many species have low levels of vitamin D and its metabolites. Both human and cow's milk contain very little vitamin D (Hillman, 1990). Store-bought cow's milk is heavily fortified with vitamin D; otherwise it would have essentially none. Human breast milk is deficient in vitamin D. Exclusively breastfed infants are at increased risk for vitamin D deficiency leading to the bone disease rickets. Current guidelines suggest supplementing breastfed infants with 400 IU vitamin D_3 per day (Wagner and Greer, 2008). Exclusively breastfed infants who do not receive vitamin D supplements have higher incidence of vitamin D insufficiency and outright deficiency (Ward et al., 2007; Gordon et al., 2008). More than 30% of vitamin D–deficient infants exhibited bone demineralization (Gordon et al., 2008) or even rachitic changes (Jain et al., 2011).

Why should milk be deficient in such a necessary nutrient? And how did our ancestors' infants survive and thrive when fed a deficient diet? The answers are: sunlight and adipose tissue.

Vitamin D is only a true vitamin for certain species (e.g., cats and polar bears; box 5.2) or under certain circumstances (e.g., when no access to direct sunlight is available). For most species, vitamin D is produced by a photosynthetic reaction in skin. Specific wavelengths of ultraviolet light

(270–300 nm; part of the UV-B spectrum) transform 7-hydrocholesterol found in skin into pre-vitamin D. Pre-vitamin D converts to vitamin D in the skin at body temperature, and the vitamin D then is absorbed into the bloodstream (see box 5.1 for more detail). Thus, in the presence of sufficient UV-B light in those specific wavelengths (wavelengths that are poorly absorbed by the atmosphere and so are present in normal sunlight), humans and most other species produce their own vitamin D and have no need for a dietary source. Some carnivores appear to have lost this ability (box 5.2), but the loss does them no harm since their natural diets always contain sufficient vitamin D (truly a vitamin in this case) from the skin and liver of their prey.

Over the course of our evolution, vitamin D was mostly acquired through sunlight. Our ancestors' infants were generally exposed to plenty of natural sunlight and thus produced plenty of vitamin D in their skin. They were self-sufficient and did not need a dietary source, which explains why humans did not evolve physiological mechanisms to concentrate vitamin D metabolites in milk. In our past, vitamin D deficiency and its associated diseases (e.g., rickets) would have been largely absent. Vitamin D deficiency is a disease of modern humanity.

Nowadays, people tend to stay out of the sun or put on sunscreens when they are exposed to sunlight. There are valid health reasons for this change. The same wavelengths of light that produce vitamin D can also damage skin and cause sunburn. Excess sun exposure is a leading risk factor for skin cancer, one of the deadliest cancers. Protecting babies from sunburn is good preventative care for their long-term health. Too much protection, however, increases the risk of vitamin D insufficiency, especially while they are being exclusively breastfed. Luckily, it takes fairly little sun exposure to produce sufficient vitamin D (box 5.1). Also, babies are usually born with enough vitamin D to last them several months, which brings us to why adipose tissue protects breastfed babies from vitamin D deficiency.

Adipose tissue acts as a storage depot for vitamin D metabolites and other fat-soluble molecules. This is one reason why vitamin D deficiency takes a long time to develop, and why humans far from the equator can remain vitamin D sufficient through the winter even with low dietary intake of vitamin D. High sun exposure during the summer months would result in several months' supply of vitamin D and its metabolites being stored in body fat. This mechanism is potentially important for infant health as well. Human

BOX 5.2

Cats and dogs don't make vitamin D

In 1918 the British doctor Edward Mellanby demonstrated that feeding dogs on a diet of porridge resulted in rickets, which could then be cured if the dogs were fed cod liver oil (Mellanby, 1918). There was some confusion at the time over what exactly was the anti-rachitic factor in cod liver oil. The American researchers Elmer McCollum and Marguerite Davis had previously discovered an essential factor extracted from cod liver oil and named it vitamin A. Mellanby suggested that either vitamin A was the anti-rachitic factor that cured his dogs, or that there was another, as yet not identified, essential factor in cod liver oil. In 1922 Elmer McCollum successfully treated rachitic rats with cod liver oil in which vitamin A had been destroyed by oxidation, demonstrating the existence of a separate anti-rachitic substance (McCollum et al., 1922). As this was the fourth such substance with vitamin activity to have been discovered it was named vitamin D.

Many mammals are capable of producing vitamin D in their skin when exposed to UV-B light (see box 5.1). This ability has been demonstrated in sheep, cattle, horses, pigs, rats, and people. It was assumed to be true for all mammals, meaning that vitamin D is not truly a vitamin. However, in the 1980s researchers demonstrated that dogs on a vitamin D–free diet developed vitamin D deficiency even if they were exposed to UV-B light (Hazewinkel et al., 1987). Follow-up experiments demonstrated that the concentration of 7-hydrocholesterol, the substance converted to previtamin D by UV-B light, was very low in the skin of both dogs and cats, about 10 times lower than in the skin of rats (How et al., 1995). Irradiation by UV-B light did not change the concentration of vitamin D in the skin of dogs and cats, but increased vitamin D concentration by 40 times in *(continued)*

infants are born with an exceptionally high amount of fat compared to many other mammalian species (Kuzawa, 1998). Vitamin D is transferred across the placenta during gestation and much of it ends up stored in adipose tissue. Even though breast milk is deficient in vitamin D (Hillman, 1990) a solely breastfed infant has enough stored vitamin D to avoid frank deficiency for months. And, of course, if the infant is exposed to enough sunlight (or another source of UV-B radiation), endogenous photosynthetic production will be sufficient.

Nonetheless, vitamin D deficiency does occur among breastfed infants. Almost one of four full-term, exclusively breastfed infants in New Zealand was found to have serum concentrations of 25(OH) vitamin D (the vitamin D

BOX 5.2 CON'T

rat skin (How et al., 1994). Kittens fed a vitamin D–free diet showed a consistent decline in 25 hydroxyvitamin D (25(OH)D) status whether they were exposed to natural sunlight, artificial UV-B light, or kept indoors (Morris et al., 1999). Since the earlier work of Mellanby and McCollum had definitively demonstrated that dietary vitamin D cured rickets in dogs, the implication was that for dogs and cats, vitamin D is truly a vitamin.

Interestingly, if kittens are given an inhibitor of the enzyme 7-dehydrocholesterol reductase, which breaks down 7-dehydrocholesterol to cholesterol, exposure to UV-B light results in a progressive increase in circulating 25(OH)D despite the lack of vitamin D in the diet (Morris et al., 1999). Feeding kittens the inhibitor resulted in a five-fold increase in skin 7-hydrocholesterol concentration. Thus, cats have all the necessary metabolic machinery to synthesize vitamin D in their skin, but they apparently convert their 7-hydrocholesterol back into cholesterol at such a rate that little if any is available for photoconversion to previtamin D.

There is evidence that polar bears also do not produce vitamin D in their skin and thus have a dietary requirement (Kenny et al., 1999). What do all these animals that need to ingest vitamin D have in common? They are all carnivores. Most mammalian livers contain large stores of both vitamin D and its first metabolite 25(OH)D. Obligate carnivores such as cats and polar bears, as well as the canids, normally eat plenty of liver, and have throughout their evolutionary history. There was no advantage to producing their own vitamin D. Based on the amount of vitamin D generally found in seal liver, polar bears would be more in danger of overdosing on vitamin D rather than ever facing any deficiency. So the ability has been lost. Cats simply convert their previtamin D back to cholesterol, which they can use in metabolism, trading a no-longer useful molecule for a useful one.

metabolite that best represents vitamin D status), which is deemed insufficient (Wall et al., 2013). Breastfed infants in Iowa were at risk of vitamin D deficiency unless they were supplemented (Ziegler et al., 2006). Marginal vitamin D status and deficiency has been reported in breastfed infants from multiple studies in South Asia (India, Pakistan, and Bangladesh), despite the lower latitude that would potentially increase UV-B exposure. In a study of term breastfed infants in India, two of three were deemed vitamin D deficient, and 30% had radiological evidence of rickets (Jain et al., 2011). In both the New Zealand and Iowa studies, infants born in the summer had higher circulating 25(OH) vitamin D than those born in the winter, probably reflecting the low levels of UV-B radiation that penetrate the atmosphere during winter. The Indian infants showed no seasonal difference in serum 25(OH) vitamin D concentration, though the mothers had higher circulating 25(OH) vitamin D in summer (Jain et al., 2011), suggesting that these infants had little exposure to sunlight in either season. Other studies of south Asian infants have detected a seasonal difference in serum 25(OH) vitamin D (e.g., Atiq et al., 1998), and one study (Agarwal et al., 2002) found that infants in an area with high air pollution had lower circulating 25(OH) vitamin D than did comparable infants in a low air pollution area. A study of Bangladesh infants between one and six months of age found that younger infants had low circulating 25(OH) vitamin D (Roth et al., 2010). The authors hypothesized that low maternal vitamin D status (due to low vitamin D diets and little sun exposure owing to cultural norms disposing women to be mostly covered while outside) resulted in poor neonatal stores, but that sun exposure increased infant circulating 25(OH) vitamin D. All of these studies are consistent with sun exposure being a critical component for maternal, neonatal, and infant vitamin D status, and that breast milk is a poor source of dietary vitamin D.

Rickets is a disease of modern civilization and of technology. We have essentially created this nutritional disease by virtue of our ability to modify our environment. Luckily there are easy solutions. Foods are often vitamin D fortified (e.g., dairy products), and oral vitamin D supplementation is effective.

Iron

Although the bioavailability of iron in milk is excellent, iron deficiency remains a concern for exclusively breastfed babies because human breast milk is low in iron. Neonates are born with a store of iron, and in humans that store is sufficient for several months. However, after that time, other iron-rich foods must be incorporated into the infant's diet, or the infant's iron status will deteriorate.

Iron deficiency is relatively rare in very young infants, at least if they were full term and not growth-restricted. Full-term neonates weighing more than 2,500 g at birth generally have substantial iron stores that were transferred across the placenta from maternal stores, and they are unlikely to develop iron insufficiency before six months of age. But if iron supplementation is not included in the diet, their risk increases with each month of being exclusively breastfed (Maguire et al., 2013). Preterm infants and term infants weighing below 2,500 g are at higher risk of iron insufficiency by six months, as their iron stores are significantly lower (Berglund et al., 2013). Maternal anemia also affects neonatal iron status, with cord blood iron values significantly lower in anemic mothers and also lower in breast milk in severely anemic mothers (Kumar et al., 2008). The incidence of iron deficiency and iron deficiency anemia in human infants varies considerably among countries, with lower incidence in developed countries where it is only a few percent. In part this reflects the higher incidence and longer length of exclusive breast-feeding in poorer countries, in conjunction with lower neonatal body stores and less availability of iron-fortified supplementary foods.

Iron in milk represents a tradeoff between adaptive characteristics. Iron is necessary for proper growth and development of young mammals. Iron also is a limiting nutrient in bacterial growth. There is a tension between providing iron to the infant through milk and reducing bacterial growth in the mammary gland and in the infant's intestine.

Milk contains lactoferrin, a molecule that binds iron. Thus, not only is the iron content of human milk low, the iron is generally bound to lactoferrin, making it unavailable for most microbes. The lactoferrin concentration in milk increases in cases of mastitis, an infection of the mammary gland (Harmon et al., 1976; Semba et al., 1999). Human milk added to the culture medium inhibits growth of *E.coli*, an effect that disappears if additional iron is added to the system saturating the milk lactoferrin (Bullen et al., 1972). Thus it would appear that human milk, while ostensibly looking like a great food for microbes (high in water and sugar) actually is a poor medium for microbial growth due, in part, to the unavailability of iron. This protects the breast from infection and lowers the microbial load delivered to infants. Although some of the microbes found in milk appear to be beneficial to the infant (see below) many others could be pathogenic, especially in high concentration. Milk iron content appears to be a compromise between infant nutritional needs and the need to protect the infant from microbial infections.

Lactoferrin may act to deliver iron to the infant as well as to sequester it from microbes. Lactoferrin binds to receptors on intestinal epithelial cells, releasing its bound iron. An evolved system to sequester iron from microbes, yet deliver it to neonates?

Maternal-Child Conflict

Successful mammalian mothers rarely only lactate once. They produce several births and nurse multiple offspring. Offspring and parents have much in common as far as fitness goals. What benefits offspring usually also benefits parental reproductive fitness. However, the most beneficial amount of resources for a parent to donate to their offspring from their fitness perspective may be less than the optimal amount for the offspring to receive from its perspective, especially in species where parents will produce many offspring over multiple births.

For this book we focus on conflict between mothers and offspring, since mammalian males don't lactate. Mothers and infants do not have identical fitness objectives. They are very similar, of course, since reproductive success by offspring is the ultimate measure of success for the mother. Fitness goals of mothers and their offspring are aligned, but they are not identical. This creates an evolutionary and selective tension, where adaptive responses that serve the mother's interests may be opposed by responses from the offspring, and vice versa. Information does not just flow from mother to offspring. Offspring can and do signal to their mothers. This is not a well-understood area of lactation science, but new evidence is suggesting that the information highway is very much a two-way street. Different kids can get different milk (see below).

Lactation is expensive—more so for some species than others—but all mothers are transferring substantial maternal resources to their offspring via milk. Most nursing females lose weight during lactation. Based on current knowledge, all nursing females lose a considerable amount of bone mineral, transferring large quantities of calcium and phosphorus from their bones to their offspring. The mother must balance her future reproductive potential against the costs and benefits of the current offspring. Or, rather, the evolutionary process has favored maternal strategies that lead to increased lifetime production of successful offspring.

In many important ways mother's milk is the perfect food for a baby. Evolution has acted so that what a female produces for her neonate is the substance that has led to the greatest fitness for that species. That is not to say that milk is in every way perfect for the neonate. Overall mean fitness of the species has been maximized, but not necessarily all components that contribute to fitness. There are likely tradeoffs between what would be best for the neonate now and what is best for the mother over the rest of her reproductive life span. Also, there can be tradeoffs between selective pressures. The main objective of this chapter is to drive home the point that milk is an evolved substance and, as such, is the result of many, sometimes conflicting, selective pressures. The current composition of a species milk, including potential variation, reflects the integration of past reproductive success within the species lineage. It is unlikely ever to be perfect; but it does represent success.

Different Milks for Different Kids

Milk is not uniform. Certainly not across mammals, but also not within a species, and it even can differ for individual females both across lactation and in different lactations. Some of this variation in milk composition reflects evolved species traits. For example, Asian elephant milk changes in composition over the first two years of lactation, decreasing in sugar and water content and increasing in fat and protein content (Abbondanza et al., 2013). The milk a calf drinks at one year of age has twice the energy per gram as the milk it drank a few weeks after birth. In contrast, many (but not all) primate milks seem not to vary in composition after the colostrum stage. Common marmoset milk shows no systematic changes in composition between 10 days and 65 days of lactation (infants are generally weaned by 70 days). However, there is substantial variation in milk composition among females and even within a female for different birth litters. Marmoset milks can range in energy content from 0.5 kcal/g to over 1 kcal/g, with sugar providing most of the energy in low-energy milks and fat providing the majority of calories in high-energy milks (Power et al., 2008). The variation in milk composition appears more related to maternal condition than to infant age or any other infant characteristic. The result, however, is that different marmoset infants may be reared on milks of different compositions. So far we have not been able to detect any consistent differences in outcomes due to variation in milk composition in this species, possibly because many other important milk constituents (e.g., protein and minerals content) adjust appropriately with the changes in fat and sugar content (see chapter 6). However, marmoset infants are vulnerable to developing obesity and obesity-related disease at an early age in captivity (Power et al., 2012; 2013). Differences in percent body fat between normal and obese one-year-old marmosets are detectable by one month of age, while they are still reliant on mother's milk for almost all of their nutrition (Power et al., 2012). We are investigating the possibility that differences in milk bioactive factors may contribute to this difference in growth and development in infant marmosets.

Amazing Marsupials

Kangaroos and wallabies provide a fascinating example of how a mother produces different milk for different kids. The pouch young, or joeys, are born after a short gestation (typically less than a month). The fetus-like joey climbs to the pouch and attaches to a nipple. The mammary gland for that nipple produces the appropriate milk for the newborn joey; high in sugars, with many oligosaccharides besides lactose, and low in fat and protein. Milk composition changes quite dramatically over time as the joey grows and matures, decreasing in sugar content and increasing in fat and protein. Before the joey is weaned, a sibling will be born and will crawl up to the pouch and attach to another nipple. Thus, a kangaroo or wallaby mother will be producing two different kinds of milk at the same time (Lincoln and Renfree, 1981; Nicholas 1988). The two mammary glands work independently of each other, producing milk appropriate for the age of the joey that feeds from them.

Although this evolved adaptive pattern of mammary development and function shows great flexibility by being able to both change milk composition over lactation and produce different milks from different mammary glands, the specific composition of the milk a mammary gland produces appears to be fixed by when nursing began. It does not appear to be able to change based on signals from the joey. A study in which tammar wallaby joeys were removed from nipples and replaced with younger joeys did not find any significant changes in the pattern of change in composition over time for the milk. In other words, the younger joeys received milk appropriate for an older joey, and not what they would have received if they had stayed on their own mother's nipple. Interestingly, the young joeys fed older joey milk grew and developed faster than controls left on their mothers (Trott et al., 2003). The fostered joeys that were 56 days younger than the joeys they replaced achieved a series of developmental markers at earlier ages than did controls. The appearance of whiskers, eyes opening, ability to stand, and having fur all over the body all occurred 35 days earlier in the fostered joeys. Fostered joeys could hop by age 172 days as opposed to 214 days in controls. By age 214 days fostered joeys had the appearance of a 242-day-old control joey. The fostered joeys also appeared fatter than controls, even at comparable developmental stages. Thus, joey age did not affect milk composition, but milk composition affected joey growth and development.

This result reinforces that milk has evolved to balance growth with development. The composition of a species' milk may not have evolved to maximize growth of the offspring, but rather to support appropriate growth. This is an especially important concept for human babies and human health; one which we develop further in later chapters when we discuss milk and the paradigm of the Developmental Origins of Heath and Disease.

Sons versus Daughters

There are intriguing data from Katie Hinde's research on milk of rhesus monkeys that imply that the composition of mother's milk is affected by the sex of the infant being nursed. Macaque milk actually changes in composition over lactation, with milks produced at three months being higher in energy content than milks produced at one month. There was no difference in milk composition in regard to infant sex at one month of age, but at three months mothers of male infants produced milk that was greater in energy density (e.g., relatively higher in fat and lower in sugar) than did mothers of female infants (Hinde, 2009). Interestingly, the calcium content of milk from mothers of female infants was higher (Hinde et al., 2013). Male rhesus macaques grow faster, but female rhesus macaques reach skeletal maturity at an earlier age. It would appear that rhesus monkey mothers may adjust the characteristics of their milk to support the differences in growth and metabolic needs of their infants, at least in regards to differences between male and female infants.

A difference in the amount of milk produced, but not the nutrient composition, between female and male offspring appears to be true for cows (Hinde et al., 2014). A dairy cow that has a daughter will produce more milk than one who gives birth to a son. This effect is especially pronounced for the first calf a cow produces. If her first-born is a daughter, a cow will have greater milk production throughout her life. This phenomenon appears to be an epigenetic development of the cow's udders, where either the hormonal milieu of a daughter enhances mammary development or the hormonal milieu of a son has a negative effect. It is not clear that this phenomenon would have an effect on the calf, however. Dairy cows have been bred to produce large quantities of milk. The birth of a daughter may mean the cow can produce more milk, but it isn't likely that her daughter would have benefited (or that her son would have suffered a deficit in the opposite case) if the cow was allowed to rear her calf. A dairy cow likely produces far more milk than any calf would need, or even be able to ingest. This phenomenon is something the dairy industry might well be interested in. The industry would prefer more female calves anyway, as male dairy cattle are not of much economic worth. If rigging births to favor daughters over sons also increases milk production that increases the incentive for mechanisms to increase the ratio of female-to-male calves. But it is not clear that this phenomenon has any adaptive significance for wild ruminants. As long as the milk production by a cow that first gives birth to a son is sufficient to rear her offspring, then the fact that another cow that gave birth to a daughter first may have a higher maximum for milk production may be evolutionarily irrelevant. The second cow may never achieve her maximum milk production under normal (not being milked by machines) circumstances.

Evolution versus Perfection

Milk composition can be considered a maternal phenotype. The phenotype includes whether the mother has any ability to adjust milk composition depending upon her circumstances or on signals or characteristics of her offspring. Like any phenotype, the genetic underpinnings are subject to selective pressures. Milk composition is the end result of millions of years of evolution and represents a successful strategy for past challenges. There are some species for which milk composition appears remarkably fixed, with little variation between females or across lactation. In those species an argument can be made that milk composition has been constrained by selective forces, and maybe it is the best it can be. Other species are capable of producing milks with a wide range of any particular constituent; most commonly a wide variation in fat content. That implies that there is no single best milk composition, or that perhaps it depends on circumstances. In all cases, milk composition is the result of competing selective pressures. What is best for the mom may not be best for the infant, or what is best for the current infant may negatively affect the mom's next infant. Tradeoffs have occurred.

Finally, milk serves multiple functions in regulating and supporting the growth and development of the neonate. Nutrition must be balanced by immune function and developmental signaling. What is "best" is a complex, multi-dimensional problem that evolution may not be able to solve for the simple reason that there may be no single "best."

Chapter 6

The Milk Spectrum

Despite our emphasis in this book on the signaling and regulatory aspects of milk, a fundamental function of lactation remains to deliver nutrients to a female's offspring. Milk provides the nutrients for metabolism, growth, and development. The macronutrients in milk are water, fat (lipid), protein, sugar, and minerals such as calcium and phosphorus. Although these ingredients are common to all milks, the relative proportions can vary tremendously among species. For example, milks of most species are composed largely of water; as much as 90–91% of rhinoceros or horse milk is water. Human milk generally is 88–89% water. Even in animals with what are thought of as high-fat milks, water is still the largest constituent. Water is an important (though often overlooked) nutrient. We will die of thirst long before we will starve to death, and that is definitely true of our babies. But some seal milks have less than 50% water, and have more fat than water! These seal milks appear very different from milk from a primate or perissodactyl (e.g., horse, rhinoceros). But these seal mammary secretions are milk nonetheless, and they derive from the same ancient ancestors as the milks of horses and people. The vast variety of mammalian species has resulted in an equally impressive variety of milks.

The milk spectrum, as we term it in this book, is the range of relative proportions of fat, sugar, and protein that are found in milks from the vast array of placental mammal species. These principal constituents of milk are usually consistently interrelated to each other in the milk of any given species; but, importantly, they can vary in absolute and relative amounts widely between species. The fact that these constituents are usually correlated stems from the obvious observation that they make up most of the milk solids. If one varies, then the others almost have to. To demonstrate, consider a bowl of heavy cream. If we add sugar to the cream (for example, in order to make ice cream) we also will have changed the relative proportions of water, fat, and protein in the mixture. Adding sugar means that in every gram of sweetened cream there has to be less of the other constituents compared to the original cream. Heavy cream you buy in the store contains 60% water, 35.9% fat, 2.3% protein, and 1.8% sugar (and other things like minerals that make up less than 1% of the cream). If we add 10 g of sugar to 100 g of cream, the composition becomes 54.5% water, 32.6% fat, 2.1% protein, and 10.7% sugar. The point is that change in one constituent results in changes to the others—at least in a bowl.

We chose the sweetened cream example because it does not exist in nature. In fact, it cannot, and the reason why helps explain one pattern of the milk spectrum. We can add sugar to cream in a bowl and nothing else will happen. Mother's milk, however, is produced in the mammary gland, and as such is surrounded by extracellular fluid. Milk and blood are isosmotic (Holt, 1983). Adding sugar into milk within a mammary gland will create an osmotic gradient and will draw water into the gland and into the milk. More accurately, sugar (lactose) produced in the Golgi apparatus draws water into the Golgi apparatus to then be transferred into the mammary gland ducts along with the other milk constituents (see chapter 3). Nature does not produce high-fat, high-sugar milks for the simple reason that as sugar enters the milk it will bring water with it, diluting the milk. The biologically more realistic example is instead of adding 10 g of sugar to our cream, we add 10 g of sugar to 100 g of water, and then pour that into our bowl of cream, which would result in: 76.2% water, 17.1% fat, 1.1% protein, and 5.6% sugar. These values are still biologically unreasonable (a milk with these concentrations would not be isosmotic with blood), but demonstrate the basic phenomenon. As milk sugar goes up, so does water concentration, thus diluting the milk solids. Ice cream is made; no mammal can produce milk with such high-sugar, high-fat, and low-water content.

The first aspect of the milk spectrum, then, is the sugar-to-fat continuum. There are milks that are high in sugar and low in fat, low in sugar and high in fat, and ones that are moderate in both. The osmotic property of sugar explains why a high-sugar, high-fat milk does not exist; but why couldn't a milk be low in both sugar and fat? There is no biochemical reason why such milk could not exist. The answer lies in the primary metabolic function of fat and sugar in milk, and how that function is generally different from the primary function of milk protein. Fat and sugar provide metabolic energy for the neonate. In general, protein and minerals provide the building blocks for growth. A low-fat, low-sugar milk generally would not provide enough energy for the necessary life processes.

The categorization of milk fat and sugar as energy substrates and milk protein as a growth substrate is a gross generalization, of course. Protein certainly can be metabolized for energy. There are species of mammals that routinely use protein as an energy source. The obligate carnivores, such as the felids, are a prime example. Cats have an obligate upregulation of the enzymes that convert amino acids into glucose. If you feed a cat a high-carbohydrate, low-protein diet, the cat will likely develop a malaise and reduce its food intake (Green et al., 2008). The reason is that it will continue to turn protein (either dietary or its own body stores) into glucose, as when it was fed a low-carbohydrate diet, and thus likely will become hyperglycemic. Cats have a high protein requirement because their main source of glucose to feed their glucose-dependent organs (e.g., brain) is catabolism of amino acids (Eisert, 2011). Many of the other obligate carnivores, such as mustelids (minks, weasels, and ferrets) and obligate piscivores within the carnivore lineage (e.g., seals and sea lions) probably also are very good at turning protein into glucose. Metabolism has evolved to match diet. Some mammals primarily use carbohydrate as their main source for glucose (frugivorous species such as primates), some rely mainly on fat (polar bears and ruminants and many other herbivores that ferment fiber into fatty acids), and some use amino acid catabolism (minks and cats). Most, of course, use a mix. An unanswered question is the extent to which neonates, feeding on milk, match their mother's adult metabolic profile. Carnivore milk generally is high in protein. It is quite likely that many carnivore babies (kittens, pups, cubs, etc.) utilize milk protein as an energy source. It is quite unlikely that primate babies, including our own, catabolize milk protein to any great extent. Thus, the energy-versus-growth substrate distinction for fat, sugar, and protein in milk may be a primate-centric idea and may not apply to carnivores and some other species. Milk protein may serve as an energy source as well as a source of amino acids for tissue building in many species.

An example of a species besides the obligate carnivores in which milk protein might serve a metabolic energy function is the giant anteater (*Myrmecophaga tridactyla*; figure 6.1). A native of Central and South America, the giant anteater is a member of the mammalian superorder Xenarthra, all of whose members (anteaters, sloths, and armadillos) share a South American origin. There is very limited data on milk from any of the Xenarthrans; but what there is suggests that protein is higher in concentration than either fat or sugar. For the giant anteater mean values are: protein = 5.8%; sugar = 3.4%;

Figure 6.1. Mother giant anteater (*Myrmecophaga tridactyla*) carrying her baby on her back. Photo: Mehgan Murphy, Smithsonian National Zoological Park, Washington DC.

fat = 1.4%. Giant anteater milk appears to be an example of a moderately low-sugar, low-fat milk. On a gross energy (GE; box 6.1) basis protein contributes 57% of the GE, sugar 22%, and fat only 21%. If all the protein, sugar, and fat in the milk were metabolized into energy (ME), protein would contribute 52% of ME, sugar 29%, and fat only 19%.

Not all the milk protein will be metabolized, of course. Some will still be used to build tissue. But the high proportion of potential metabolizable energy in giant anteater milk from protein strongly suggests that protein is an important energy source for the neonates. This is also probably true for adults; a diet of ants is likely high in protein, moderate at best in fat, and low in carbohydrate (as well as high in dirt). Catabolism of protein into glucose would be predicted to be important for adult giant anteaters.

A reliance on milk protein for energy may be a feature of Xenarthran lactation. Armadillo milk is also high in protein and low in sugar, though it is not low in fat. Milk protein accounts for about 50% of GE in armadillo milk. Adult armadillo diet also is high in protein, and probably low in fat and carbohydrate. There are no data on sloth milk as yet, but sloths have a diet high in mature leaves, which typically are high in protein, and low in

BOX 6.1

Gross energy (GE) of milk

The gross energy (GE) content of any substance represents the total energy in the substance if released by combustion. GE is measured directly by bomb calorimetry, in which a dried sample of a substance is combusted in a high-oxygen environment. Milk GE can be calculated from its fat, protein, and sugar content (measured in grams per gram of milk) by the following formula:

$$GE = 3.95 \times sugar + 9.11 \times fat + 5.86 \times protein$$

This formula has been shown to produce GE values for milks of Asian elephant, rhesus macaque, Weddell seal, and rhinoceros that are indistinguishable from values measured directly by bomb calorimetry (Abbondanza et al., 2013).

GE is always greater than the metabolizable energy (ME), which is the energy that can be extracted from ingested food by an animal, primarily due to the inability of animals to completely metabolize protein. Metabolism of amino acids results in nitrogenous waste products that contain approximately 30% of the GE of the amino acid. The rough formula for ME of any food is:

$$ME = 4 \times sugar + 9 \times fat + 4 \times protein$$

fat and simple sugars. The extant Xenarthran might share a milk spectrum that reflects milk protein as an important energy source. Further research is needed to determine the extent to which this hypothesis reflects evolutionary reality.

Armadillo milk provides an example of a further complication concerning protein content of milk. The calcium and phosphorus in milk is primarily bound up in the casein micelles. Armadillo pups have an unusual growth challenge in that they must grow the bony plates that serve as armadillo armor. Not surprisingly, armadillo milk is very high in total mineral content, with mineral content greater than 3% (which exceeds sugar content) in the latter half of lactation (figure 6.2) and calcium content exceeding 1% at the end of lactation. Armadillo milk calcium content is strongly associated with milk protein content (figure 6.3), suggesting a high ratio of casein proteins to whey proteins. The high protein content of armadillo milk may be required for the growth of bony plates in order to transfer the appropriate amounts of calcium and phosphorus from mother to offspring via milk. The digestion of the casein proteins releases the minerals but also provides a large amount

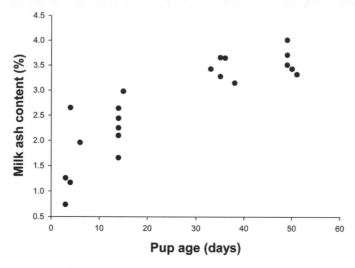

Figure 6.2. Mineral content of nine-banded armadillo milk across lactation.

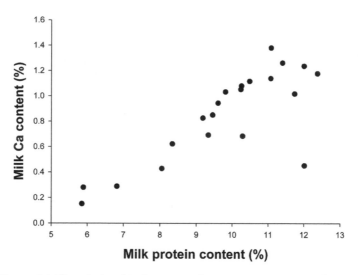

Figure 6.3. The relationship between calcium concentration and protein in armadillo milk.

of amino acids for metabolism. Armadillo milk protein content exceeds 10% during late lactation and probably provides a substantial excess of amino acids relative to the growth requirement. The excess amino acids will be catabolized for metabolic energy. A high protein milk is required for a high calcium-phosphorus milk and also would favor the utilization of protein for energy by pups. We suggest that an enhanced ability to use protein as an energy substrate may be an ancestral trait in Xenarthrans and is associated with milk protein being an important energy substrate for offspring. The high milk protein content may have been a preadaptation that allowed the evolution of the bony plates in armadillos by enabling an associated high milk mineral content.

The other species of mammal that has armor-like plates is the pangolin or scaly anteater. Pangolin scales are composed of keratin, not bone; however, the keratinized scales are high in the sulfur amino acid cysteine, which provides the disulfide bridges that contribute to the strength and rigidity of keratin. Pangolins are closely related to carnivores, which have high-protein milks, in general. We predict that pangolin milk will have a high relative protein content, but with a substantial whey component to provide cysteine (casein proteins are deficient in sulfur amino acids). Pangolin adults and pups likely have a good ability to use protein as an energy substrate. An enhanced ability to metabolize protein for energy leading to a high protein milk may have served as a preadaptation to the evolution of the pangolin scale.

We suggest that the evolution of armadillo bony plates and pangolin keratinized scales required a milk relatively high in protein content. In the case of armadillo milk, we predict the milk protein to be high in casein proteins and for pangolin milk, high in whey proteins. In both cases we predict that pups will utilize milk protein for energy in addition to depositing it into tissue for growth.

Nothing Is Simple for Milk

While it is defensible and even productive to consider milk composition in terms of crude fat, crude protein, and total sugar (and those are the parameters most commonly measured in the lab), in reality milk is made up of fats, proteins, and sugars. In other words, these three nutrient categories are comprised of multiple biochemical entities that can have very different biological effects. Many fats (e.g., short- and medium-chain fatty acids) serve predominantly as an energy substrate. Long-chain fatty acids certainly can be used for energy; but they are also important constituents in cell membranes and in metabolic pathways. Not all milk fat is used for energy; some is deposited into tissue. Similarly, not all milk sugar is metabolized. Lactose, the predominant sugar in many but not all milks, is an energy substrate. But milk contains other oligosaccharides that do not appear to be metabolized, at least by placental mammals. These sugars do not provide energy to the neonate, but do serve anti- and prebiotic functions. Milk oligosaccharides may also be incorporated into tissue, though that has not been demonstrated. The protein fraction of milk includes immunoglobulins, hormones, and other bioactive peptides that serve functions outside of nutrition.

So the simple formulation that milk fat and sugar provide energy and milk protein supports growth must give way to the more accurate observation that not all fats and sugars provide energy, and some protein undoubtedly does, and some fats and maybe sugars are required for appropriate growth. The extent to which the functions of fat, sugar, and protein deviate from the simplistic formulation varies by taxa. For many carnivore babies, milk sugar probably provides minimal energy, while milk protein represents a significant source of metabolizable energy in the milk; for primate babies, the reverse is probably true.

Lactation Stage

When we are discussing the milk spectrum we are talking about the composition of mature milk; the milk at mid-lactation (often termed peak lactation). Milk very early or late in lactation can differ quite remarkably from mature milk, for very different reasons. Milk produced at the end of lactation is usually lower in water and sugar and higher in fat and protein. This arises from the simple fact that lactogenesis (synthesis of lactose) is becoming downregulated in the mammary gland. At the end of lactation the mammary gland undergoes a process called involution, which involves apoptosis (programmed cell death) and extensive tissue remodeling. The mammary epithelial cells responsible for turning the mammary gland into a milk-synthesizing machine postpartum are removed from the gland. In some species this process is very rapid; in others it occurs over an extended period of time. In the latter species, milk composition gradually changes, becoming lower in sugar and water and higher in fat and protein. The change is driven by the osmotic gradient between milk and blood; less lactose, lower gradient, less water drawn into the milk to match the milk with blood. That appears as an increase in protein and fat, but is more biologically understandable as a decrease in water and sugar. Whether this change in nutrient composition is meaningful for the offspring is unknown. The offspring is self-feeding at this time, but still might obtain critical nutrition from milk. Some authors have suggested an adaptive purpose to this compositional change, but it may just represent the biochemical consequences of the loss of lactose-producing cells.

Milk at the very beginning of lactation, within the first few days of birth (colostrum), is high in protein and can be low in both sugar and fat. Colostrum serves a developmental and immunological function more than a nutritional function. Most of the protein in colostrum represents immunoglobulins as well as bioactive peptides that have function in development including, but not restricted to, the immune system. The primary adaptive purpose of colostrum differs from that of mature milk. The high protein content may partly reflect the low lactose and fat synthesis in early lactation as much as a heightened protein synthesis. Protein synthesis (of immunoglobulins and peptides, but not necessarily of the nutrient-carrying casein proteins) precedes lactose and fat synthesis. For example, mammary secretions can be collected from many species before birth. In the white rhinoceros, mature milk contains about 8% sugar, 1% protein, and 0.5% fat, and thus about 90% water. Mammary secretions from a white rhinoceros a few days before the calf is born contains less than 2% sugar, essentially no fat, and about 24% protein, leaving only 74% water. It looks more like proteinaceous saliva than milk. Perhaps these prebirth secretions are reminiscent of the early synapsid proto milk discussed in chapters 2 and 3?

The high protein content of this rhinoceros prepartum mammary secretion (it really isn't milk yet) certainly represents a high rate of immunoglobulin synthesis, as would be expected; it also represents a low rate of lactose synthesis, and thus a low osmotic gradient resulting in less water entering the mammary gland. It probably also represents the fact that the secretions are accumulating in the gland (no suckling is occurring, obviously). A store of immunoglobulins is accumulating, to be passed on to the rhinoceros calf soon after birth. If the secretions were being consistently removed from the mammary the protein content would probably appear lower. Still, within the first three months after birth, before fat and sugar content has achieved mature milk concentration, rhinoceros milk protein content is somewhat high at 1.4%. After three months postpartum it is steady at 1%. Thus, in the white rhinoceros protein synthesis is higher for an extended period in early lactation.

Milk Fat

Fat content is the most variable nutrient constituent across mammals. The milks of some species contain almost no fat (e.g., white rhinoceros, ring-tailed lemur) while the milks of some aquatic mammals contain as much or more fat as they do water (e.g., hooded seals with 50–60% fat). Fat is an excellent energy source to power metabolism; it contains more than twice the metabolizable energy per gram than either sugar or protein. High-fat milk allows a large amount of energy to be transferred to the neonate in a small volume. High-sugar milk would require a much larger volume for the same amount of energy. In chapter 7 we discuss some of the advantages and disadvantages of fat over sugar as an energy source in milk, and why a species might follow a high-fat lactation strategy versus a high-sugar lactation strategy.

The process by which lipid is secreted into milk appears to be unique to mammary gland cells and differs from the exocytosis used by other cells to secrete lipid (Heid and Keenan, 2005). The fat in milk is highly structured. It is packaged in a milk fat globule which is surrounded by a milk fat globule membrane. The milk fat globule is a spherical structure with the triglyceride center surrounded by phospholipids and proteins, especially glycoproteins, forming the membrane, and with a glycosylated surface. The milk fat globule membrane originates from the plasma membrane of mammary gland secretory cells. Microlipid droplets composed of triglycerides coated by proteins and polar lipids are released from endoplasmic reticulum into the cytosol where they fuse to form larger cytoplasmic lipid droplets. These droplets are transported to apical cell regions where they are enveloped by apical plasma membrane. The droplets push out until the enclosing plasma membrane fuses at the base, completely enveloping the milk fat globule and closing off the apical cell membrane.

Milk fat globules can range in diameter from 0.2 microns (10^{-6} m) to as large as 15 microns. It is not known whether the range in milk fat globule size results from random processes or is somehow regulated (Mather and Keenan, 1998). Milk fat globule diameter is one factor determining how naturally homogenized milk is, with a smaller diameter associated with a better dispersion of milk fat globules in the milk, and a lower tendency to form a cream layer over time. For example, goat milk fat globules on average have a diameter of 2.76 microns, while cow milk fat globules are on average 3.51 microns in diameter (Attaie and Richter, 2000), providing one explanation for why goat milk is considered to be "naturally homogenized." The homogenization process for cow milk reduces mean fat globule size to 0.4 microns, much smaller than goat milk fat globule diameter, indicating that size alone does not explain the difference in "creaming" between cow and goat milk. Although a smaller milk fat globule is one reason why goat milk separates into a cream layer more slowly than cow milk, a more important reason is that cow milk contains a protein that enhances the agglutination of milk fat globules to form larger milk fat structures. Goat milk lacks this protein. In a sense, it is still about size; the large, agglutinated clusters of milk fat globules in nonhomogenized cow milk are more likely to rise to the top of a glass of milk.

Many proteins are incorporated into the milk fat globule membrane (Mather, 2000; Pisanu et al., 2011). Some of these proteins are simply carryovers from the proteins in the apical cell membrane, but others appear vitally necessary for the formation of the milk fat globule membrane. The proteins butyrophilin and xanthine oxidoreductase appear vital to milk fat globule membrane formation. Knockout mice lacking either of these two genes fail to produce milk with the normal fat content (Ogg et al., 2004; Vorbach et al., 2006). In the case of xanthine oxidoreductase, its lack causes a failure of the milk fat globule to be enveloped by the apical epithelial plasma membrane, resulting in a rupture of the mammary epithelium resulting in premature involution of the mammary gland. It seems that milk fat needs to be packaged.

Some of the milk fat globule membrane proteins may perform important metabolic functions after digestion. At the least they provide amino acids. But some may have regulatory functions as well (discussed in chapter 9). Many also have antimicrobial and pro- and anti-inflammatory functions in other tissues. In addition to its vital role in milk fat globule membrane formation, xanthine oxidoreductase has antimicrobial and antiviral actions. The importance of xanthine oxidoreductase to the innate immune system as well as its necessity for milk fat globule formation is a major factor in suggesting that early proto lactation arose from skin glands whose secretions supported antimicrobial functions related to the innate immune system and the inflammatory response (Vorbach et al., 2006). The ancestral skin secretions likely served to protect maternal skin in addition to any function directed at eggs/offspring.

Sugar

All milks contain some sugar, generally in the form of oligosaccharides, short polymers of simple sugars typically consisting of two to ten simple sugars bound together (box 6.2). For many mammals oligosaccharides are the most common milk constituent, after water. In most milks the oligosaccharide lactose is the predominant sugar; this is especially true of the high-sugar milks (5–9% sugar) such as milks produced by primates and perissodactyls (e.g., horses and rhinoceros). But milk from all species contains some oligosaccharides other than lactose. Other oligosaccharides besides lactose are predominant in the early milks of many marsupials, though we regretfully will not be able to consider marsupial milk composition in detail in this book. Milks from marine mammals generally contain very little if any lactose, though these milks typically contain up to about 1% sugar. Most or all of the sugar in these milks is in the form of other, larger, more complex oligosaccharides. Human milk is unusual among primates in that as much as 25% of the sugar in human milk consists of oligosaccharides other than lactose, with the balance being lactose and a small amount of monosaccharides such as glucose and galactose.

The ability to synthesize lactose is an ancient mammalian adaptation tied to lactation, and it requires the peptide α-lactalbumin (see chapter 3). But certain modern mammalian taxa have lost the ability to express a functional α-lactalbumin molecule and thus cannot synthesize lactose. Two branches of the pinniped radiation, the otariids (fur seals and sea lions) and the odobenids (walrus) have mutations of the α-lactalbumin gene that appear to render it non-functional, at least as far as being a co-enzyme for producing lactose (Reich and Arnould, 2007). In contrast, the α-lactalbumin gene sequenced from several phocid species (true seals) showed high sequence identity with dog α-lactalbumin, and would appear to be completely functional. Those species still produce milk with low sugar content; but the milk does contain lactose. In chapter 7 we discuss why losing the ability to produce lactose is a successful adaptation for otariids.

Lactose provides metabolizable energy for mammalian neonates because young mammals express the intestinal enzyme lactase. Adult mammals rarely express this enzyme and thus are lactose intolerant. Certain groups of human beings are exceptions to this general rule, expressing lactase into adulthood and thus able to digest lactose efficiently (see chapter 11). Lactose is a more effective way to provide sugar to offspring, due to the greater

BOX 6.2

Milk oligosaccharides

Sugar, or more accurately, carbohydrates, come in many forms. Simple sugars, such as glucose, galactose, and fructose, are monosaccharides. Disaccharides consist of two monosaccharaides bound together. For example, sucrose (table sugar) is composed of a glucose molecule bound to a fructose molecule; maltose, a disaccharide found in grains such as barley, is composed of two molecules of glucose; and milk sugar, or lactose, is composed of a molecule of glucose and a molecule of galactose. Oligosaccharide is the general name for a sugar molecule made up of two or more monosaccharides; thus, disaccharides are a special type of oligosaccharide. Lactose is an oligosaccharide, but it is generally excluded from the discussion of milk oligosaccharides because the biological function of lactose is clear. Infants express the enzyme lactase in their small intestine, which cleaves the bond between the glucose and galactose molecules, producing two monosaccharides that can be easily absorbed and metabolized. Lactose is a principal metabolic energy source for most (but not all) mammalian infants.

Milk from all species so far examined contains other, larger, more complex oligosaccharides, whose functions are just beginning to be understood. For the milks of most species, lactose is the overwhelmingly predominant sugar, with milk oligosaccharides comprising only a small proportion of the sugar fraction of the milk. Many carnivore species and marine mammals, however, produce milks with very little sugar, in some cases less than 1%, and the sugars in these milks are predominantly milk oligosaccharides. Primates produce relatively high-sugar milks in which lactose is predominant and other oligosaccharides comprise a small fraction of the milk. Human milk is exceptional among primate milks in that milk oligosaccharides comprise about 25% of the sugar in human milk, with the rest being lactose and small quantities of glucose and galactose.

Oligosaccharides coat the surface of cells of all living animals, as well as many circulating proteins through conjugation to serine, threonine, or asparagine. Oligosaccharides are also found conjugated to cell surface lipids. As such, oligosaccharides are involved in a host of processes including, but not limited to, cellular recognition, cell adhesion, fertilization, and immune function.

osmotic properties relative to energy density of monosaccharides such as glucose, as discussed in chapter 3. If we are considering the neonatal nutrition of a primate or a perissodactyl (e.g., horse, rhinoceros) then lactose is an extremely important milk nutrient that contributes substantially to metabolic energy. For some carnivores lactose may be almost irrelevant, and it is non-existent in the milk of fur seals and sea lions. But all milks do contain some sugar, just as all contain some fat and protein.

Food for Bugs

The other milk oligosaccharides do not appear to have a nutritional function, at least for the placental mammal infant. They do not appear to be able to be digested and absorbed (this may not be true for marsupial neonates, which may metabolize oligosaccharides). Some milk oligosaccharides perform an antimicrobial function. We discuss this function of milk oligosaccharides in greater detail in chapter 8. Some of the oligosaccharides in milk appear to act as prebiotics. Certain commensal gut microbes (e.g., *Bifidobacterium infantis*) are able to metabolize specific human milk oligosaccharides, while many pathogenic strains cannot (Ward et al., 2006; Sela and Mills, 2010). Milk is feeding more than the infant; it is providing food for the gut microbes as well.

This certainly benefits the infant. An appropriate gut microbiome is important for health and well-being. It effects immune function, digestive function, and serves as a defense against potential pathogenic microbes. Mother's milk and the infant gut microbiome appear to have coevolved (see chapter 8 for further discussion).

Protein

There are three basic categories of milk proteins: caseins, whey proteins, and milk fat globule membrane proteins. Most milk proteins are digested and provide a source of amino acids for the infant. They may have other functions within the mammary gland, within the milk, and possibly before they are digested by the infant, but in the end most serve as food. Nonetheless, many proteins are resistant to digestion and are able to perform other functions within the infant's digestive tract.

There are binding proteins and enzymes in milk that are resistant to proteolysis in the gut (e.g., lactoferrin, bile salt stimulated lipase (BSSL), and, in human milk at least, amylase). The iron-binding peptide lactoferrin can be found intact in the stools of neonates (Davidson and Lönnerdal, 1987) and can bind to receptors found on cells in the small intestine of neonates (Kawakami and Lönnerdal, 1991; Suzuki et al., 2001), possibly serving to deliver iron to the neonate. Milk provides the neonate with some of the enzymes needed to digest milk. The digestion of the lipid from the milk fat globule is accomplished in part by BSSL, which is in the milk. Neonates produce low levels of lipases and other digestive enzymes. Heat treating human milk decreases fat absorption in neonates, indicating that enzymes in milk, such as BSSL, are important for milk digestion (Andersson et al., 2007). Curiously, amylase, a starch-digesting enzyme, is found in human and chimpanzee milk. Its possible functions are discussed in chapter 12.

Many of the proteins in milk appear to have conserved regions within them, which suggest that these fragments may have biological activity after digestion. The entire protein has little or perhaps some particular function, but selection has acted to produce and conserve cleavage sites such that bio-active fragments will be released upon digestion. This is a major area of current research to identify these potentially important peptides hidden within larger milk proteins. We discuss this in more detail in chapter 9.

Milk proteins provide necessary nutrition. Protein (and minerals) in milk can be viewed as a major constraint on growth, at least for lean body mass growth. The higher the protein concentration in a species' milk, the faster, in general, its babies will grow. But what constitutes high-protein milk? And how consistent is the relationship between milk protein content and neonatal growth patterns? For many taxa, milk proteins provide energy as well as the building blocks for tissue, complicating the association of protein with lean body mass growth. Another major hurdle to examining the relationship between milk protein content and growth is the sugar-to-fat milk spectrum. High-sugar milks will look low in protein compared to low-sugar milks; but the biological significance is not clear. The low protein content of many high-sugar milks may simply reflect a dilution effect of water.

Milk protein concentration can be easily measured in the laboratory. The tricky issue is how to express milk protein content in a biologically meaningful way so that high-sugar and low-sugar milks can be compared. We suggest examining nutrient content in relationship to milk energy.

Energy

Energy is a fundamental factor and constraint in living systems. Energy is required to combat the consequences of the second law of thermodynamics, which states, in simplistic terms, that the order of a system must decrease over time (entropy, or disorder, must increase). Living cells and living beings represent an increase in order relative to non-living objects. An energy input is required for organisms to create the more-ordered conditions necessary for life. For plants, the ultimate source of that energy is sunlight. For animals, food is eaten, to be broken down by digestion, which releases energy by increasing the disorder of the food. Metabolism then shuttles that energy within the body to create and maintain the ordered systems required for life. Energy is a fundamental input for living things, just as waste products are a fundamental output.

The energy in milk derives from the milk fats, carbohydrates, and proteins. For the purposes of the final section in this chapter we will assume that the vast majority of metabolic energy from milk derives from the fats and carbohydrates, and little from the proteins. For many species this is a reasonable assumption; but as mentioned above, it likely fails for others. For humans, and other primates, it is probably a useful and largely correct approximation to the metabolic functions of these milk constituents. Based on this assumption we divide the energy in milk into protein energy and non-protein energy. The total energy in milk (termed the gross energy, or GE; box 6.1) is the sum of non-protein and protein energy.

In general, milks can be characterized along a continuum from milks where almost all the non-protein energy comes from fat to milks where most of the non-protein energy comes from sugar (mostly from lactose). We believe characterizing milk by the contributions to milk energy from its major constituents has advantages over expressing milk constituents by their concentrations. The fat, sugar, and protein content of milks can be viewed graphically by plotting the percent of total energy (milk GE) from sugar against the percent of milk GE from fat (figure 6.4). The percent of milk GE from protein is just 100% minus the sum of the percentages of GE from fat and sugar; but it can be seen graphically as the 45 degree lines (slope of −1) that connect the sugar and fat axes. Most mammalian milks fall in a range of 20–60% of GE from protein (figure 6.4), with the energy from fat and sugar ranging from less than 10% to around 75% for sugar and 95% for fat.

How to Compare Milks

There are many solutions to the challenge of feeding a baby mammal to weaning. Many successful milks have evolved whose composition vary substantially, primarily across the fat-sugar spectrum, but also in protein and minerals. Comparing milks that differ substantially on the fat-sugar continuum can be especially difficult. High-sugar milks will be low in fat and

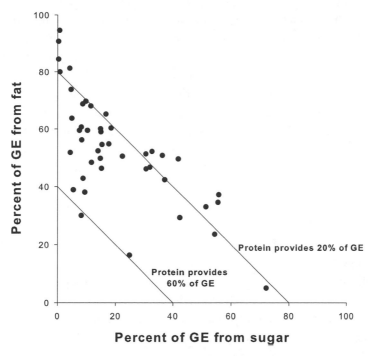

Figure 6.4. Percent of GE from fat versus percent of GE from sugar for many mammal milks. The two points closest to the line of 60% GE from protein are the Xenarthrans, the nine-banded armadillo and the giant anteater. The point next closest to that line is the African lion.

protein, in part due to their high water content. Low sugar milks generally will be the reverse. These differences are important and have biological consequences and meaning, but sometimes they are difficult to interpret. For example, what is the best way to compare milk protein content? Milks vary extensively in protein concentration among mammals. But milk protein concentration needs to be examined in relation to the other milk constituents. The relative proportions of protein may be more important than the absolute concentration.

Milks of humans, the great apes, horses, and rhinoceroses have protein contents around 1%. Whole cow milk you buy in the store has 3% protein. The milk from a Weddell seal has almost 16% protein. At first glance these milks would appear easy to put in order of increasing protein; but not if we consider milk protein in relation to the other constituents. Human milk contains almost 4% fat, cow milk 3% fat, the great apes 2–3% fat, and horses and rhinoceroses only around 1% fat. Weddell seal milk has more than 50% fat (Eisert et al., 2013)! Weddell seal milk has 17 times the fat of store-bought whole cow milk, but only a bit over 5 times the protein. Human milk has the same protein content as milk of the white rhinoceros, but more than 4 times the fat content. One way to resolve the difficulty of comparing milks with different levels of all main constituents is to express the constituents on an energy basis.

From the perspective of a nutritionist formulating a diet, the concentration of a nutrient on an energy basis is a fundamental unit. It derives from the concept that energy requirement is the primary driving force in ingestion. A formulated diet is considered nutritionally appropriate if when the animal consumes a sufficient amount of the diet to meet its energy needs it also consumes appropriate amounts of all its required nutrients. Milk can be considered to be the whole diet of neonates. We suggest that evaluating milk as a diet, rather than as a food, can be illuminating.

We can apply this concept to milk by expressing the major constituents (fat, sugar, and protein) on an energy basis, as we did graphically in figure 6.4. This method of expressing milk constituents uses what is called the Geometric Framework (Raubenheimer et al., 2009). The Geometric Framework, a methodology for evaluating the dietary ecology of species, can provide a useful means of comparing milks in a biologically meaningful way, and may be especially valuable in relating milk composition to the growth patterns of the neonates.

Our first example concerns the milks produced by Asian elephants and white rhinoceroses. Both of these mammals are large, terrestrial herbivores that produce a single large calf that grows relatively rapidly at about 1kg per day. Elephants are browsers, feeding on trees, shrubs, vines, and so forth. Rhinos are grazers, feeding primarily on grasses. The milks that Asian elephant and white rhinoceros cows produce appear to be quite different. Asian elephant milk composition at mid-lactation (between 4 and 7 months of calf age) contains about 4.9% sugar, 3.6% protein, 8.3% fat, and about 80% water (Abbondanza et al., 2013), while the milk of a white rhinoceros at 3 months contains 6.9% sugar, 1.1% protein, 0.5% fat, and over 90% water (Power, unpublished data). Asian elephant milk changes in composition consistently as the calf ages over the first 18 months of life, increasing in fat to about 15% and protein to over 5%, while sugar decreases to less than 4% (figure 6.5). White rhino milk composition remains constant after 3 months of calf age.

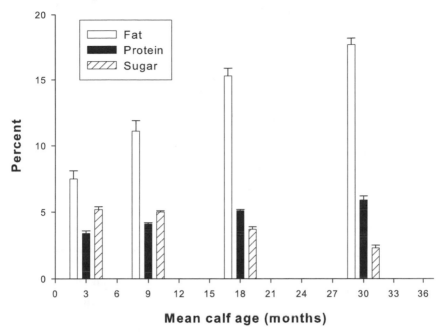

Figure 6.5. Mean values for the nutrient composition of Asian elephant milk from the first month of life through 30 months. From Abbondanza et al., 2013; reproduced with permission.

These two species have evolved different lactation strategies that solve the same basic problem—growing a large calf relatively quickly.

However, if we examine milk protein on an energy basis, we find that despite the obvious essential differences between elephant milk (high-fat, high-protein, low-sugar) and rhinoceros milk (low-fat, low-protein, high-sugar), the amount of protein per non-protein energy in these milks are extremely similar. Milk protein accounts for about 17% of milk GE for both species. This similarity in milk protein matches the similarity in growth rates for elephant and rhinoceros calves. Thus, elephants and rhinos demonstrate that there are multiple lactation strategies that can solve the problem of producing a fast-growing, large, terrestrial herbivore; but the apparent difference in milk protein content may not be biologically meaningful. It may reflect a difference in how we measure milk protein content in the laboratory, but not a metabolic difference in a neonate. In at least one sense (percent of energy from protein) their milks are more similar than they are different.

To examine this concept in more detail, we consider the changes in milk composition as the Asian elephant calf ages. When we plot the percent of

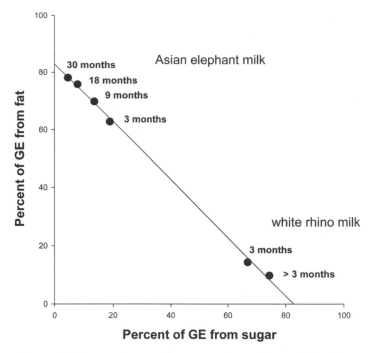

Figure 6.6. Percent of energy from sugar, fat, and protein for Asian elephant milk and white rhinoceros milk.

GE in milk from sugar on the horizontal axis and the percent of GE from fat on the vertical axis, we see in figure 6.6 that Asian elephant milk composition does change over time, with the percent of GE from sugar decreasing and the percent of GE from fat increasing with calf age. However, the sum of the percent of GE from sugar and the percent of GE from fat remains constant, at about 82–83% of GE. In other words, the percent of GE from protein has not changed. All the elephant milk points are in the top-left quadrant of the graph, and move up and to the left with calf age along the line of equal protein as a percent of GE (17%). In the lower-right quadrant of the graph are the points representing white rhinoceros milk composition at before 3 months of calf age and after. Quite different from elephant milk, of course, in the percentages of GE from sugar and fat; but falling quite close to the line describing the percent of GE from protein for elephant milk. Elephant milk and rhinoceros milk are basically equivalent protein sources.

In the Asian elephant, the average protein content of milk steadily increases as the calf ages as measured by the percent of protein, but so does the energy content. Although the absolute protein concentration in the milk can range from about 3.5% soon after birth to more than 5% after one year, the amount of protein in 1 kcal of milk is remarkably constant at about 30mg. Of course 1 kcal of milk would weigh about 1g during the first few months, but would only weigh about 0.6 g after the calf is one year old, due to the increase in milk energy driven by the increase in milk fat. Thus, for an elephant cow to transfer 1,000 kcal (1 million calories) to her calf it would take about 1,000 g (about1 liter) of milk in the first few months of life, but only 600 g (0.6 liters) after the calf's first birthday. Regardless, the mother would be providing her calf with about 30 g of protein at both times. Milk protein in relation to energy appears to be stable even though the fat and sugar content (and hence energy content) of milk is changing.

Due to her lower-energy, dilute milk, a white rhinoceros cow would have to feed her calf with more than 2.6 liters of her milk in order to provide 1,000 kcal. But 2.6 liters of rhinoceros milk would contain approximately the same amount of protein as 1,000 kcal of elephant milk (estimated at 29 g). Elephant and rhinoceros milk are different; but as protein sources they are roughly equivalent.

The biggest difference between elephant and rhinoceros milk is the water content. A rhinoceros cow transfers much more water to her calf for every kcal than does an elephant cow. Assuming that all milk protein is deposited into tissue, based on the growth rates of white rhinoceros calves we calculated that a cow must deliver about 40 kg of milk per day to her calf, of which 36 kg is water. That amount of milk would represent over 15,000 kcal/day of total milk energy for the calf (they are big and they grow fast). For an Asian elephant cow to transfer the same amount of energy and protein would only require 15 kg of milk when her calf was newborn, and 9 kg of milk at 18 months, representing a transfer of 12 or 7.5 kg of water, respectively. Elephant and rhinoceros milk may be equivalent protein sources, but their differing lactation strategies have potentially significant consequences for maternal and calf water balance.

A final comment on milk protein and calf growth in the white rhinoceros: before three months postpartum white rhino milk contains relatively more energy from fat and protein than after 3 months (figure 6.6). The difference in protein energy is not substantial, but the point for before

3 months is below the line of 17% energy from protein and the point for after 3 months is above the line. The closer to the origin the higher the percent of energy from protein (100% energy from protein would be the origin). After three months of age, a white rhinoceros cow produces a milk with 16–17% energy from protein; between 1 and 3 months (after colostrum) the milk contains about 18% energy from protein. White rhino calves grow fastest in the first three months of life, consistent with our hypothesis that milk protein expressed on a per-energy basis is associated with growth rate.

Different Milks Solve Different Problems

We hope the reader is beginning to get a sense of why milk composition varies so much across different mammalian species. The examples in this chapter just begin to scratch the surface of selective pressures that have shaped the milks of mammals. In the next chapter we expand on this discussion to explore lactation strategies, the integration of milk composition, milk volume, nursing frequency, and duration of lactation.

Chapter 7

Lactation Strategies

Infants of different mammal species vary tremendously in their initial state of development at birth, their rate of growth and development, their energy expenditure, and thus their nutrient requirements. The marsupial mammals produce fetal-like neonates that attach to nipples in the pouch and grow and develop fueled solely by milk. The placental mammals have actually simplified lactation by shifting embryonic and fetal nutrition and regulatory signaling to the placenta. But even among placental mammals there is great diversity in the state of development of offspring at birth. There are precocial species, such as horses and giraffes, in which the offspring are born so well developed they can walk and follow their mothers within hours of birth. Dolphins and whales produce extremely precocial offspring which must be able to swim to the surface of the water immediately for their first breath. At the other end of the developmental continuum are the altricial species, whose offspring are born poorly developed. Although altricial whales don't exist, as they would have all drowned at birth, there are many altricial mammal species, such as mice, cats, and dogs, in which babies are born essentially helpless, with eyes and ears closed. Bears represent an extreme for placental mammals, producing fetal-like cubs.

Offspring birth condition is only one aspect that affects the nutrient needs of the offspring and the maternal ability to meet those needs. Other factors include rate of growth and development, maternal ecology and reproductive effort, and the eventual weaning diet for the offspring. It is not surprising that, although all mammals lactate, the specifics of lactation vary extensively across mammalian taxa. To accompany the great diversity among mammals in these parameters, a remarkable number of lactation strategies have evolved, all with their own brand of milk.

What Is a Lactation Strategy?

The lactation strategy of a species is the integration of the duration of lactation, the frequency of suckling, the volume of milk the mother produces per

BOX 7.1

Life History Strategy

A species's life history strategy refers to the suite of evolved characteristics of a species from birth through adulthood that determines reproductive success. It incorporates the obvious concept that to be reproductively successful an individual has to succeed at every life stage. The focus has often been on reproductive characteristics such as the number of offspring produced in each reproductive effort, the size of offspring, their rate of growth and development, the frequency of breeding, and the age distribution of reproductive effort (e.g., time to first reproduction and total reproductive lifetime). Life history theory came out of ecology. One of its first manifestations was the concept of *r*- versus *k*-selected species in which, in its simplest form, r-selected species produce many offspring over a short life span as opposed to k-selected species that produce a few offspring at a time over a long reproductive life span. The theory has now been extended to almost all aspects of biological study, including evolutionary psychology (e.g., Wolf et al., 2007; Stearns et al., 2008). More recently, genomics has begun to be explicitly incorporated (Roff, 2007). There are many books on life history theory/strategy dating from the beginning of the field (e.g., Stearns, 1992; Roff, 1993) to more recent edited volumes that explore the diversity of research in this area (e.g., Flatt and Heyland, 2011).

We argue that for mammals, the lactation strategy is an integral part of a species's life history strategy, affected by and affecting most other aspects of life history.

day, the number of offspring being nursed, and the composition of that milk. All of these components must be considered together to understand how a species is meeting the challenge of providing its offspring with adequate nutrition. Any single component by itself, while informative, does not provide the complete answer. All of these parameters interact with the growth rate, developmental pattern, and digestive and metabolic capabilities of the offspring. The lactation strategy must also be compatible with the ecology and reproductive strategy of the mother. The lactation strategy of a mammalian species is central to its life history strategy (box 7.1). In this chapter we give examples of different lactation strategies and explore how they "fit" the requirements and biology of the species. How fast do the offspring grow? How many are there? Are they nursed frequently with only a short interval between nursing bouts, or are nursing bouts rare and separated by a substantial amount of time? How soon are offspring weaned? How frequently will the mother reproduce, and does she need to end lactation before she becomes fertile again? What foods are available for offspring at weaning? All these factors and more influence the lactation strategy of a species. The main focus of this chapter is how different milk compositions fit into different lactation strategies. All aspects must meld together, but our interest is how the other aspects of a lactation strategy affect milk composition.

Some Milks Change, Others Stay the Same

There are at least two different biologically important factors that underlie changes in milk composition: (1) consistent changes over lactation as the neonate(s) grow and develop; and (2) changes in milk composition driven by maternal differences. The first type of compositional change is important for understanding the differences between species. The second type of change is important for understanding differences between individuals of a species. Both can operate within a species. For example, in the Asian elephant, different cows produced milks with significantly different fat and sugar contents; but in all cases the milk also changed consistently with calf age, becoming higher in fat and lower in sugar as the calf got older (Abbondanza et al., 2013). The milk an Asian elephant calf received was determined both by something intrinsic to its mother and also by the calf's age.

All milks change in composition over time. The milk produced right after birth (colostrum) does not serve the same purpose as milk later in a neonate's life. The difference between colostrum and mature milk has already been discussed in chapter 6. In this section we are concerned with mature milk and whether it changes in composition or stays stable. For many species, once the mammary begins to produce what is termed "mature milk" the composition changes little if at all until the end of lactation. For other species, however, milk composition continually changes, often quite dramatically, as the offspring age. These patterns represent different lactation strategies.

In some species maternal condition can affect milk composition; in other species milk is more invariant. For example, among primates, several lemur species from Madagascar produce relatively similar and invariant milks. It appears impossible to distinguish between the milk of different females, or even between species. Other primate species appear to produce more variable milks. There is convincing evidence in the rhesus macaque and the common marmoset that female condition has a significant effect on milk

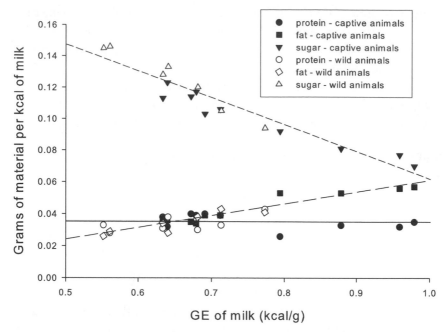

Figure 7.1. Fat, sugar, and protein content of common marmoset milk on a per energy basis (g/kcal). Data from Power et al., 2008.

composition. Even in these variable milks, however, there appears to be aspects of constancy, which may reflect phylogenetic or infant developmental constraints.

In the common marmoset, a small New World monkey, milk nutrient composition not only varied among females but also could be different between births for the same female. However, milk composition for any particular infant was reasonably constant across lactation, with no systematic changes with infant age (Power et al., 2008). In marmoset milk, the nutrients that varied between females or between lactation events for the same female were fat and sugar. Fat is the milk constituent that displays the greatest variation, ranging from under 1% to as high as 13%. The protein content of marmoset milk is remarkably constant, when expressed on a per-energy basis. Female marmosets that produce low-energy milk provide more mg of sugar for every kcal of milk and fewer mg of fat per kcal of milk than females that produce high energy milk; but all females provide about 35 mg/kcal protein in their milk (figure 7.1).

The relative contributions to milk energy from sugar and fat were associated with maternal body condition and, more specifically, change in body mass during lactation (Tardif et al., 2001). In general, larger females produced higher-fat milks and a greater proportion of energy from fat. The mean fat content of milks from normal-sized dams was significantly less than the milk fat content of milks from large dams. As a result, the energy substrate from milk to support infant metabolism differs depending on maternal condition. Infants of large dams receive almost half (48.9%) of their total milk energy intake from fat, and fat accounts for 61% of the non-protein energy. Infants of normal-sized dams receive about equal amounts of milk energy from fat and carbohydrate (about 40% of total milk energy). Milk protein accounts for about 21% of energy and does not vary across dams.

Among normal-sized marmoset dams, body mass change over lactation affected milk composition. Females that lost weight during lactation produced higher-sugar, lower-fat milk compared to females that maintained or gained weight (Tardif et al., 2001). Since large females were less likely to lose weight during lactation, this effect of body mass change on milk composition may explain, at least in part, the high-fat milks of large marmoset females. Once again, there was no effect on milk protein when expressed on an energy basis. A female that loses weight will provide more calories to her infants from sugar and less from fat, but will provide the same amount of calories from protein.

The stability of milk composition across lactation probably depends primarily on infant development patterns, though constraints on the mother may also play a role in some species. Species that produce altricial young would be predicted to display greater changes in milk composition across lactation, as neonatal nutrient requirements are more likely to differ from those of older unweaned offspring in those species just due to the different developmental states. The marsupials are the poster children for that effect. Precocial infants are predicted to have more stable requirements over lactation and thus be able to thrive on milk that varies less at different ages. But even infants of precocial species can have variable growth and developmental patterns with age. The more stable the pattern of growth, the more likely a milk of relatively consistent composition will suffice.

It is important to remember that there may be multiple lactation strategies that could be successfully employed to solve almost any combination of neonatal and maternal constraints and challenges. Comparing lactation strategies between species can be like comparing apples and oranges. Phylogenetic constraints may play as large a role as other factors. It is possible that the patterns of milk composition for a species were largely set in the ancient ancestor of their mammalian lineage, and that milk composition has constrained the infant developmental patterns and other aspects of life history of a species. A primate might never be able to evolve a milk composition similar to that of a carnivore, even if that primate became more carnivorous.

Length of Lactation

The length of time that mammalian mothers' nurse their offspring varies remarkably between species. Not surprisingly, the length of lactation has an effect on milk composition, though its effects are moderated by factors such as whether neonates are born altricial or precocial, and at what developmental stage offspring are self-feeding.

Hooded seals hold the record for the shortest lactation period at less than four days (Oftedal et al., 1993a). Gestation is quite long in hooded seals, and the pups are the most precocial of the pinnipeds. Hooded seal milk contains high concentrations of fat, exceeding 50% and peaking at around 60%. Hooded seal pups gain a remarkable amount of mass in a short time, doubling their birth weight over four days. Their mass gain is mainly the result of storing fat from their mother's milk; they don't truly grow until that fat is metabolized. Mass gain is not synonymous with growth, and growth is not synonymous with development. In the case of the hooded seal, the initial mass gain is an unregulated process driven simply by the extensive quantities of fat (and protein and minerals) transferred from the mother to her pup over an extremely short time period (4 days). True growth and development then occurs over the succeeding months while the pup fasts (mom has left, never to return). This remarkable adaptive strategy shows an extreme end to the range of possible successful lactation strategies.

An almost opposite lactation strategy is followed by the great apes (orangutans, gorillas, and chimpanzees). Ape infants are born precocial, in that their eyes and ears are open, and they can grasp and cling onto their mothers. Ape babies can be considered altricial if they are compared to a giraffe calf or dolphin calf; but compared to a hooded seal pup there isn't any obvious behavioral developmental difference. An ape neonate clings to its mother; a hooded seal pup lies on pack ice. Neither can effectively locomote on its own.

An ape baby will be nursed for 4 to 5 years. Milk supplies essentially all nutrition for most of the first year of life, and the majority of nutrition well into the second year. Ape milk is about as different in composition from hooded seal milk as you can get, with 7–8% sugar, 2–3% fat, 1% protein, and 88% water.

A cautious generalization is that species with a short lactation period will produce low-sugar, high-fat, high-protein milk. Species with long lactations are more likely to produce a high-sugar milk. The shorter the lactation, the more energy and protein that must be delivered each day, suggesting the need for a more concentrated (i.e., low-water-content) milk. Other factors can modify this proposed association. The closer two species are to each other phylogenetically, in body size, and in ecology the more likely this association will hold. Plenty of counter examples exist; for example, Asian elephants and white rhinoceroses, discussed in chapter 6. Rhinoceros cows produce high-sugar, low-fat milk; elephants produce milk higher in fat and lower in sugar. Weaning in the white rhinoceros occurs at about one year; in elephants, cows may nurse their calves for 5–10 years. The higher fat content of elephant milk is not in response to a short lactation.

Nursing Frequency

The association between nursing frequency and milk composition is stronger than the one between composition and the length of lactation. Species that nurse frequently are more likely to produce high-sugar milk. Again, phylogeny, body size, infant development pattern, and other factors exert their own influences that can reinforce or oppose the effect of nursing frequency. But a species whose breeding females nurse their offspring infrequently are predicted to produce concentrated milk (low-sugar) while a species whose females nurse their young frequently can manage with a high-sugar, and thus dilute, milk. Milk composition appears largely consistent with nursing patterns.

For example, rabbits leave their kits in a nest or burrow. They visit them once per day to nurse them. One adaptive purpose of this restricted nursing behavior is that it lessens the chance that predators can find the nest by observing the coming and going of the doe. Tree shrews go rabbits one step further. Tree shrew mothers park their babies in a special nest, completely separate from the nest the mother uses. The female tree shrew returns to her babies every 48 hours to nurse them for a few minutes at most. It is estimated that total nursing time is about 25 minutes over about four weeks, after which the pups are fully weaned (Martin, 1966).

Most primate mothers carry their babies with them and nurse them frequently. Human mother's nurse their babies about every 2–4 hours. We consider primate lactation strategies in more detail at the end of this chapter. In brief, anthropoid primates generally produce dilute, high-sugar milks.

As you can imagine, the milks of rabbits and tree shrews contain much higher concentrations of solids, with high levels of fat and protein, compared to the milks from apes and humans, who carry their infants with them and nurse every few hours. Volume constraints on both the maternal mammary gland and the neonatal stomach provide the explanation for concentrated milks with infrequent nursing. A female tree shrew must be able to deliver two days' worth of nutrition to her pups (litters range from one to three pups); the pups' stomachs must be able to hold that much milk.

In contrast, a mother monkey of about the same size (e.g., common marmoset) will deliver approximately the same amount of total nutrition over 48 hours in many nursing bouts, probably more than 12 and maybe as many as 20. Each marmoset nursing bout would deliver a fraction of the energy and protein, but much more water than the single tree shrew nursing bout.

Water in milk serves an important purpose; death by dehydration comes much more quickly than death by starvation. Neonates are vulnerable to dehydration. Marmoset infants are not likely to suffer from a lack of water, even though they are exposed to the elements (heat, sunlight, air), because they receive a steady supply via milk. Tree shrew babies would be more vulnerable to dehydration, but they are safely within a mostly enclosed nest, which likely reduces water loss. The milk composition suitable for the marmoset lactation strategy would fail for tree shrews. It couldn't deliver enough nutrition in a single nursing bout every 48 hours. The milk composition suitable for the tree shrew strategy would make little sense for a marmoset; and might expose infants to the risk of dehydration.

In the next sections we examine the lactation strategies of a few mammalian taxa to see how their milk composition fits into the rest of their life history strategy. We end with a discussion of primates, in order to set up our discussion of the evolution of human milk in chapter 12.

Lactating While Fasting

Many bear species have a unique challenge to their reproduction. They give birth and initially nurse their cubs while they are hibernating. This means they are lactating while they are fasting. Milk composition is reliant on body stores and metabolism alone.

Bears produce extremely altricial cubs, perhaps the most altricial of any placental mammal. Black bear neonates weigh less than 0.4% of maternal weight (Oftedal et al., 1993b). Black bear sows den up approximately 6–8 weeks before giving birth and do not leave the den until their cubs are about 3 months old. Up to 5 months of growth of the fetus and then the cub(s) must be supported by black bear sows without any food or water intake.

Black bear milk is high in fat (20–30%) and low in sugar (1–3%) and water (60–70%). Fat and protein increase over lactation while sugar remains constant. Relatively little of the sugar in black bear milk appears to be lactose; most probably black bear milk sugar is comprised of larger oligosaccharides (Oftedal et al., 1993b), which may serve more of a pre-biotic or anti-biotic function than a nutritional one.

Why do black bears produce altricial cubs which they feed on a high-fat milk? The answer relates to glucose and possibly water conservation. The mammalian brain runs on glucose. Brains cannot easily employ fatty acid metabolism. During starvation the brain can exist on ketones, but it is not the preferred substrate. The placental-fetal unit also preferentially utilizes glucose for energy. During mammalian pregnancy, insulin resistance increases, reducing glucose utilization by muscle tissue, an adaptive physiological response to conserve glucose for the fetus and for the maternal brain. Fasting during pregnancy challenges this evolved metabolic system. At some point competition between the maternal brain and the placental-fetal unit for glucose will challenge the mother's physiological capacity for gluconeogenesis. The bear solution is to remove the placental-fetal unit from the mother (i.e., give birth to an altricial cub) and begin to deliver energy to the neonate through the cub's intestinal tract. Fat from body stores can be mobilized and given to the cub directly, sparing maternal glucose.

Water conservation also undoubtedly plays a role. The mother has no access to water. The cubs require water. Black bears are large animals, and mother bears can go quite a while without water before they become dehydrated, especially while hibernating, which lowers both energy and water loss. But milk fed to her cubs represents water loss. The lower the water content of her milk, the less water she loses. A low-sugar milk serves both a glucose-sparing and a water-sparing function.

Nursing her cubs still presents a fluid balance challenge to the sow. Low water content is still 60–70% water. Black bear sows appear to be good at recycling. The milk fed to their cubs goes through the digestive tract, providing nutrients for the cubs, but also producing feces and urine. The evidence is clear that the black bear sow reingests the excreta from her cubs, recycling as much of the water she gave her cubs as possible (Oftedal et al., 1993b).

Interestingly, even bears that do not hibernate (e.g., pandas) produce altricial cubs and a low-sugar milk. It appears that phylogenetic constraints act to conserve the bear reproductive pattern. The hypothesis is that the ancestral bear was a hibernating animal. As its descendants spread toward warmer climes they gave up hibernation, but apparently could not change the pattern of producing an altricial cub. Milk composition for non-hibernating bears is similar to that of hibernating bears. For example, giant pandas produce a low-sugar milk (about 1–2%). Milk composition that is necessary for a fasting bear appears also to be successful for non-fasting bears.

The giant panda has evolved a successful dietary ecology in which they eat large quantities of bamboo which they digest poorly, assimilating perhaps as little as 20% of the available energy (Dierenfeld et al., 1982). It is possible that retaining the bear reproductive strategy of producing an extremely altricial cub and feeding it a low-sugar, high-fat milk allowed the ancestors of giant pandas to diverge from an omnivorous animal to become an herbivorous bamboo specialist. The giant panda exhibits exceptionally low daily energy expenditure, living on less than 40% of the energy expected for a mammal of its body mass (Nie et al., 2015). They achieve this low energy expenditure by having smaller-than-expected brains, livers, and kidneys for their body size, low levels of physical activity compared with other bears, a thick pelage that reduces heat loss, and low levels of thyroid hormones, which probably indicates low basal metabolic rate (Nie et al., 2015). Indeed, their thyroid hormone levels are lower than those found in hibernating black bears. A panda probably faces similar constraints and challenges to its glucose metabolism during gestation as does a hibernating black bear. Pandas may not hibernate, but their everyday low rate of energy intake presents a challenge. Glucose, which will be in short supply, needs to be partitioned between maternal brain and the fetal-placental unit. A reproductive and lactation strategy appropriate for a hibernating bear may have been a necessary preadaptation for giant pandas to evolve to become bamboo specialists.

Fasting Seals

Another mammalian taxa that lactates while fasting are the phocid seals (family Phocidae). All phocid seal species give birth on land (often on ice), which means a phocid seal cow nurses her pup away from where she feeds, presenting a conflict between feeding herself and feeding her pup. Lactation is generally short, ranging from the incredibly short, less than 4-day lactation of the hooded seal to just over one month in the ringed seal (Lydersen and Kovacs, 1999) and the Weddell seal (Eisert et al., 2013). Phocid seal milk is the highest in fat concentration, routinely exceeding 50% fat. Phocid cows transfer a large amount of nutritional resources to their pups in a short time. Phocid pups are born precocial, with eyes and ears open. However, they are generally sedentary, spending a majority of their time asleep, and thus expend little energy (Lydersen and Kovacs, 1999). The seals with the shortest lactations (e.g., hooded seals [4 days] and harp seals [12 days]) fast through the entire lactation. Weddell seals that lactate for about 40 days do leave their pups to feed after the first few weeks (Eisert and Oftedal, 2009), but cows still lose up to 40% of their body mass during lactation (Eisert et al., 2013). These seals are large animals; Weddell seals are around 1,000 pounds. Large size is a necessary condition for this kind of lactation strategy.

Large body size allows for large amounts of nutrient stores; but most nutrient stores will consist of fat and protein. Even large mammals do not have large stores of carbohydrate (e.g., glucose and glycogen) on their bodies, though the larger the mammal the greater the proportion of energy requirement it can store in glucose/glycogen. Large size is a necessary but not sufficient condition for lactating while fasting. A low-sugar milk appears also to be required, for maternal glucose and water conservation.

Sea Lions

Sea lions have a mutated form of the α-lactalbumin gene that is nonfunctional. They cannot produce lactose. They share this mutation with fur seals and walruses (Reich and Arnould, 2007). So of course all these milks are low in sugar, and any sugar in the milk is a larger oligosaccharide probably with prebiotic and antimicrobial function and little or no direct nutritional function for the pup.

Sea lions are well adapted to an aquatic existence, but they give birth and nurse their pups on land. Sea lions also have an unusual lactation strategy, unique among mammals. They differ from the phocid seals in having a long lactation, in some species lasting up to 12 months. They nurse their pups for several days and then leave to go on foraging trips that last days and even weeks. They essentially stop and restart lactation. In most mammal species, no suckling by the young for days would result in mammary gland involution and the cessation of lactation. The mammary gland would cease being a secretory gland and revert to its resting state. Sea lions maintain their mammary glands as secretory organs with no suckling while they forage. The lack of a functional α-lactalbumin gene may be an important adaptation that allows this lactation strategy. Production of lactose would draw water into the mammary glands, distending them if the milk is not expressed. If lactose synthesis occurred even at a low rate while sea lion mothers foraged away from their pups for many days, the distension of the mammary gland would lead to mammary involution and the shutting down of lactose synthesis. The mammary gland would cease to be in a secretory condition, or it would burst. But with no lactose synthesis, the volume of milk produced over the time the mother is away foraging does not place this challenge on the mammary gland.

Hiders versus Followers

Bovids (cloven-hoofed ruminants) have undergone extensive radiation producing two subfamilies (Bovinae and Antilopinae) and 12 tribes containing more than 50 genera (Groves and Grubb, 2011). Species from all but one of these tribes inhabit Africa. African bovids represent a highly diverse group with divergent neonatal-care strategies. Milk composition reflects that diversity, especially in the relative proportion of milk protein.

The African bovids can be categorized by maternal rearing / calf behavior after birth into "hiders" in which the calf stays hidden and does not follow the mother for a period of weeks to months (e.g., 2–6 weeks in dorcas gazelle and 3–7 weeks in greater kudu), or "followers" in which calves accompany their mothers within hours after birth, even during migratory movements (e.g., wildebeest) (Lent, 1974). Hider species produce milk with a higher proportion of energy from protein compared to the milk of follower species (table 7.1). Of the African bovid species in table 7.1, eland, greater kudu, dorcas gazelle, bongos, blue duikers, gemsbok, and sable antelope may be classified as "hiders" (Lent, 1974; Smithers, 1983; Ralls et al., 1986) and most have milk protein energy > 30% (mean 32.6% ± 1.6SEM; Petzinger et al., 2014). Buffalo, African indigenous cattle (two types), ibex, blesbok, and black and blue wildebeest are "followers" (Lent, 1974; Smithers, 1983; Ralls et al., 1986) and all have milk protein energy <30% and significantly lower than milk of hider species so far studied (mean 23.5% ± 1.3SEM, $p = 0.001$; Petzinger et al., 2014).

Table 7.1. Protein on a per energy basis for African bovids

Species (common name)	Tribe	Maternal care	% GE from CP	CP mg/kcal	Source
Eland	Tragelaphini	Hider*	39	66	Osthoff et al., 2012
Greater kudu	Tragelaphini	Hider	35	60	Osthoff et al., 2012
Dorcas gazelle	Antilopini	Hider	34	57	Shkolnik et al., 1980
Bongo	Tragelaphini	Hider	33	57	Petzinger et al., 2014
Blue duiker	Cephalophini	Hider	31	53	Wilson and Hirst, 1977; Taylor et al., 1990
Gemsbok	Hippotragini	Hider	30	51	Osthoff et al., 2012
Tuli cattle	Bovini	Follower†	28	48	Myburgh et al., 2012
Black wildebeest	Alcelaphini	Follower	28	47	Osthoff et al., 2009a
Sable antelope	Hippotragini	Hider	26	45	Osthoff et al., 2007b
Blesbok	Alcelaphini	Follower	25	43	Osthoff et al., 2009a
Springbok	Antilopini	Follower‡	23	39	Osthoff et al., 2007a
Nguni cattle	Bovini	Follower	21	36	Myburgh et al., 2012
Blue wildebeest	Alcelaphini	Follower	21	36	Osthoff et al., 2009a
African buffalo	Bovini	Follower	21	35	Osthoff et al., 2009b
Ibex	Caprini	Follower	21	35	Shkolnik et al., 1980

*A "hider" is a calf who stays hidden and does not follow the mother for a period of weeks to months (e.g., 2–6 weeks in dorcas gazelle and 3–7 weeks in greater kudu).

†A "follower" is a calf who accompanies its mother within hours after birth, even during migratory movements (e.g., wildebeest).

‡Springbok calves are hiders for the first few days, then become followers.

In comparing a north temperate hider (mule deer, *Odocoileus hemionus*) to a follower (mountain goat, *Oreamnos americanus*), Carl and Robbins (1988) demonstrated that the fawns of the hider species expended less energy and devoted more nutritional resources to growth in early to mid-lactation than did the follower kids; the hider fawns also ingested milk that was higher in protein relative to energy, with protein energy equal to 38% of GE in mule deer versus 32% in mountain goats (calculated from data in Oftedal and Iverson [1995]). These results provide support for the hypothesis that a higher percentage of milk energy from protein is associated with a relatively faster growth rate. It is an intuitively appealing hypothesis. Follower species produce highly precocial calves with well-developed musculature; the calves rapidly obtain coordination (or die). Follower calves must expend energy in locomotion to keep up with their mother and may be exposed to thermoregulatory and evaporative water-loss challenges. Hider calves, snuggled into a grass nest, expend little energy in locomotion and should be buffered from thermoregulatory and water-loss challenges. They are born less precocial (they can't be called altricial) and need to devote more nutritional resources into growth and development and less into other life challenges. Follower calves need more non-protein energy for metabolism and possibly more water for thermoregulation; follower calves need less metabolic energy but more protein and minerals for growth. Milk composition matches calf need.

Springbok appear to have an intermediate postnatal pattern (i.e., a short hiding phase of a few days after which they follow the herds [Smithers, 1983]). Springbok milk protein energy as a percent of GE resembles that of a follower (22.7%; table 7.1); of course the assayed milk samples were primarily from when the calf was old enough to be a follower. Samples of springbok milk from immediately after birth would be predicted to be higher in protein because of the high concentration of immunoglobulins in colostrum; but it would be of interest to test whether it would also be higher in casein protein as well, supporting rapid growth with more amino acids and calcium and phosphorus. Further research is needed on milk composition, postnatal behavior, and postnatal growth to ascertain if this apparent distinction between hiders and followers holds for additional bovid taxa and if it is correlated to hider-follower differences in postnatal energetics and growth.

Lactating During the Bad Months

In a number of species with relatively tight seasonal breeding, lactation, which is usually the most nutritionally expensive part of the female reproductive cycle, occurs during a season of low food availability. In theory this is adaptive despite the stress on lactating females because an adult female is more physically and behaviorally capable of withstanding nutritional stress than a young animal. An offspring weaned during a time of low food availability or water stress will probably die, while a lactating female and her baby have a better chance of both surviving the bad season due to the greater competence and resources of the female and the nutrition (including water) delivered to her baby via milk.

Ring-tailed lemurs (*Lemur catta*) in Madagascar lactate during the hot, dry season, where both food and water are in short supply. Lactating ring-tailed lemurs reduced activity and fed on foods with a higher water content compared to their behavior during gestation, which occurs during a cooler season (Gould et al., 2011). Weaning occurs after the hot season at the beginning of the wet season, when food availability is greater.

Ring-tailed lemurs produce milk with low fat (less than 1%), low protein (about 1.5%), and high sugar (about 8%), with a correspondingly high water content (89–90%); a dilute, low-energy milk. Ring-tailed lemurs are thought to have evolved a reproductive strategy that emphasizes a low rate of energy expenditure/transfer per day, to allow them to be income breeders (largely obtaining the required nutrients from diet, not body stores) under conditions of food limitations (Gould et al., 2011). Although the milk is low-energy, the energy from protein is not particularly low at about 20% of GE, providing about 32 mg of protein for every kcal of milk. This protein content on an energy basis is comparable to New World monkeys and higher than Old World anthropoids, implying that lemur babies are relatively fast growing, similar to New World monkeys (see last section of this chapter). Lemur milk appears well suited for milk energy prioritized for growth and not for activity, and to providing the infant with copious amounts of water, which might be crucial in their hot, dry environment.

Mother lemurs are certainly under both food and water stress during lactation; but clearly not so critically as to endanger their survival. All 12 lactating females in the Gould et al., (2011) study survived. Younger, smaller, less-competent animals might be more adversely affected by the hot, dry months with little food available. Lemur offspring do not have to face the hot season until they are about 1 year old. Lactation ends after about 5 to 6 months, during a time of relative food abundance. The young have about 6 months to further prepare for the bad season. Still, less than 50% of ring-tailed lemur babies may survive past 1 year of age in particularly harsh times (Gould et al., 2003). The lactation strategy of the ring-tailed lemur appears adapted to supporting infants at the mother's expense through the worst season lasting 3–4 months, providing the infant sufficient water for survival and a relatively high protein content to support rapid growth to maximize the probability the infant will be sufficiently developed to survive the next bad season on its own.

Allonursing

Lactation is costly for mothers. It is a necessary cost for mammalian reproductive fitness, but the general prediction would be that mothers should restrict their transfer of nutrients and other resources via milk to their own offspring, and not nurse other females' babies. However, in many mammal species allonursing, the nursing of a baby that is not the mother's own offspring, occurs at least occasionally. This behavior has been documented in bats, rodents, primates, carnivores, seals and sea lions, and ungulates (Packer et al., 1992). The underlying cause of allonursing behavior in species appears to range from female error or "milk stealing" by infants, to selective, deliberate choice by females. In some cases the allonursing is in response to loss, either of the female's offspring or of the infant's mother, but in other cases females nurse both their own offspring and the offspring of other living females. A degree of sociality is a necessary precondition for allonursing, though in a few species the sociality is limited to a site where offspring are parked. In other species there are strong social bonds among females that participate in allonursing. In a few species the behavior is so well established as to be considered communal nursing. Below we examine a few species that illustrate each type of allonursing.

Milk stealing appears to be more common in species that produce a single young (Packer et al., 1992). In hooded seals, seal pups will attempt to nurse from any female. In a few instances, after a mother has left her pup at the end of the four-day lactation period, the pup will be able to "steal" milk from another female. All hooded seal pups gain an impressive amount of weight during the short lactation period, but the pups that steal some milk gain even more mass, more than doubling their birth weight. It isn't clear if these super pups gain any lasting advantage or whether the milk stealing harms the female or her pup in the long term; but some hooded seal pups gain in the short term from stealing milk.

Not all allonursing in species with singleton offspring represents milk stealing. In many primate species it appears to be a female choice. Allonursing has been documented in many primate species, both in captivity and in the wild (Packer et al., 1992). In captive squirrel monkeys, females that have lost their infants, either through late miscarriage or early neonatal death, will occasionally nurse other infants (Williams et al., 1994). Milk from two allomothers was compared to milk from six mothers and, although the milk from the allomothers was lower in fat and hence total gross energy, the relative proportions of protein, fat, and sugar were the same as in the milk of the mothers (Milligan et al., 2008a). Importantly, the protein content on an energy basis was identical. Thus, the milk from these two females that had lost their infants was normal squirrel monkey milk, with similar benefits for the infants they nursed. It is possible that the lower milk fat reflected poorer maternal condition, and that poorer condition also contributed to the infant loss.

Female gray mouse lemur (*Microcebus murinus*) forages solitarily at night but returns to a nest that is shared among several females (2–5) to sleep during the day; these females are all from the same matriline (i.e., they share a common female ancestor). There is a relatively short breeding season restricted to approximately 4 weeks, and most females in a sleeping group will produce infants (1–3) at about the same time. The infants remain in the nest, and when a nesting site is changed all the females and their offspring move together to the new nesting site (Eberle and Kappeler, 2006). Females carry only their own offspring, but will groom and nurse any infant, though they devote most of their care toward their own (Eberle and Kappeler, 2006). In the event of maternal death, the mother's infants will be adopted and cared for by another of the females. Thus, female gray mouse lemurs are able to distinguish between their own infants and the infants of their maternal kin, but they are willing to expend effort and resources on their kin's infants, including providing milk. Mouse lemurs appear to form matrilinear cooperative care-giving reproductive units, and kin selection is a tenable hypothesis to explain this allonursing behavior.

The evening bat (*Nycticeius humeralis*) roosts in colonies of 15–300 and females give birth to twin pups during a relatively short birth season. During early lactation mothers only nurse their own pups, but when their pups are older, females will on occasion nurse pups not their own. By the time pups are about one month old 20% of observed nursing events were by a pup other than the female's offspring (Wilkinson, 1992). This does not appear to be simply a case of female error or milk stealing, as females will reject some pups and not others. Male pups were rejected more often than female pups. Females did not nurse related pups, so kin selection is not an explanation for the allonursing events. Wilkinson (1992) suggests two possible selective advantages to females that nurse another female's pup, one immediate and one longer term. The females that engaged in allonursing had been successful in recent foraging efforts, based on body weights, and might be able to produce above-average quantities of milk. Thus, nursing another pup did not deprive her own pups and may have had some advantage in maintaining milk production by emptying the mammary gland. Females would also benefit from having an empty mammary gland when leaving the roost to forage, as that will reduce her flying weight. A possible longer-term advantage is to increase the number of females in the roost in the future. The time period when allonursing occurs is during the time the bat pups are leaving the roost and engaging in independent foraging. Death by starvation is common, especially among male pups. Female pups may be buffered from starvation by the willingness of females other than their mother to nurse them on occasion. Female evening bats often forage together, and unsuccessful foragers will follow successful foragers on subsequent trips (Wilkinson,

1992). Thus, increasing the number of females in a roost could provide foraging benefits for all females.

In a few species, allonursing and communal infant care have been combined into communal nursing. This has been documented in rodents and some carnivores. Female Norway rats will nest together. If their pups are similar in age, both females will nurse all the pups, and pup survival is good. Interestingly, however, if the pups differ by two weeks or more, the younger pups do poorly, likely due to being outcompeted at the nipples (Mennella et al., 1990).

In the dwarf mongoose (*Helogale parvula*) and the meerkat (*Suricata suricatta*) a dominant female produces most if not all of the offspring, but many females in the group will lactate and nurse the pups. In the dwarf mongoose, females with no visible signs of ever being pregnant will spontaneously lactate and nurse the pups of the dominant female (Rood, 1980). These spontaneously lactating females have been shown to have experienced a pseudopregnancy (Creel et al., 1991). Based on genealogical data, the helper subordinate females are closely related to the dominant female, being largely comprised of sisters, daughters, granddaughters, and nieces, with only a few immigrant females that might not be related to the dominant female (Creel and Waser, 1994). Thus, staying with their natal group and nursing the pups of a dominant, related female provides indirect, inclusive fitness benefit.

The meerkat is similar to the dwarf mongoose in that a dominant female will produce most of the pups, while subordinate females provide the babysitting (protection) and some will lactate and nurse the dominant female's pups. Allonursing females were subordinate females in good condition that were likely genetically related to either the dominant female or the male. Energy expenditure by the dominant female, measured by the doubly labeled water technique, decreased with the number of helpers. The dominant female was able to maintain body weight through lactation, but allonursing females lost weight. Their loss in body energy per day was equivalent to the energy in about 35 g of milk, enough to support the daily energy needs of a single pup (Scantlebury et al., 2002). Thus, helpers and allonursers likely increase the litter size a dominant female meerkat can successfully raise and protect her body condition such that she can produce more pups, on average, over the year.

Primates

We are primates. We have a lactation strategy (frequent suckling, dilute milk, and a long lactation period) that is consistent with the anthropoid primate (monkeys and apes) norm. For most anthropoid primates lactation is longer than gestation. Compared to many mammalian species, primates deliver relatively little nutrition to their infants each day; but they do it over a long time.

That is not to say that there is no variation in milk composition among primates. There definitely is. Mostly the variation between taxa is in milk protein content. In some species milk fat content can be quite variable between females. There are also prosimian primates, separate from the anthropoid primates. Or perhaps a better, more scientific, distinction is that there are strepsirrhine primates (the lemurs and lorises) and the haplorhine primates (tarsiers, monkeys, and apes). In common parlance, tarsiers often are referred to as prosimians, but they are in the haplorhine clade, not the strepsirrhine clade. Sadly, the composition of tarsier milk is unknown, so we cannot examine whether the tarsier lactation phenotype is similar to the other haplorhines or is more like a prosimian.

Lactation strategy and milk composition is more variable within the strepsirrhines. Many of the lemurs, such as ring-tailed lemurs (*Lemur catta*) and *Eulemur* species, employ the same lactation strategy as the anthropoid primates, in which mothers carry their infants and nurse them frequently. These species produce high-sugar, low-fat, dilute milk. In contrast, there are species of bushbabies (the genera *Otolemur* and *Galago*) that park their babies in nests and nurse them infrequently. Species of the Madagascar lemur genus *Varecia* (the ruffed lemurs) produce litters of two or three infants and park them in a nest while the mothers' forage. These species produce milk that is lower in sugar and considerably higher in fat (Tilden and Oftedal, 1997).

The strepsirrhines provide a nice example of lactation strategy and milk composition being consistent as opposed to phylogeny ruling phenotype. Frequent nursing is associated with dilute milk; infrequent nursing with concentrated milk (Tilden and Oftedal, 1997). There is no single primate milk to which a species' lactation strategy must conform; milk composition can evolve to match changes in lactation strategy.

Monkeys versus Apes

There are three main lineages of anthropoid primates: the New World monkeys, the Old World monkeys, and the apes. Humans are members of the ape lineage. Our closest living relatives are the chimpanzees and bonobos (genus *Pan*). The general lactation strategy in all three lineages is similar, as noted above. All species carry their babies with them (though in the small New World monkeys it often will be the father, not the mother, doing the carrying). All species have relatively long lactations, relative to gestation; but that is a factor where there is variation among the lineages. The apes, in general, have significantly longer lactation periods, lasting years. In monkeys lactation is usually measured in months, ranging from 2–3 months in the small New World monkeys to as much as 9–12 months in some Old World monkeys. There is evidence of lactation in some monkeys lasting longer than 12 months; but offspring are self-feeding by then and the importance of milk is uncertain. Samples of milk from females whose offspring are older than 1 year usually display an altered composition, with lower sugar content, suggesting that lactose synthesis is declining and that lactation is slowly ending (M. L. Power, personal observation). Regardless, in the wild, the great apes (orangutans, gorillas, chimpanzees, and bonobos) do not wean their infants until 4–5 years of age. Milk can be reliably obtained from captive ape mothers 3–4 years after birth and beyond (M. L. Power, personal observation). Great ape lactation is longer than monkey lactation. Ape milks are generally more dilute, lower in fat and especially in protein than monkey milks, reflecting the slower rate of growth and development of apes relative to monkeys.

Humans are part of the great ape lineage (or they are part of ours). However, modern human lactation behavior has diverged somewhat from the pattern of the great apes. There are still many cultures where babies are nursed for 2–3 years, longer than the norm for monkeys, but a shorter duration of lactation than is found in our closest relatives. In many developed countries, average duration of lactation is much shorter. The duration of exclusive breastfeeding, even in cultures with long lactations, is often quite short, with supplemental foods given to babies within the first few months or even weeks of life. Some women can choose not to breastfeed at all. Technology has allowed humans to depart from the ancestral lactation strategy. We can produce foods that infants can ingest and digest. A wild baby chimpanzee or gorilla does not have that option; accordingly, they typically do not ingest much if any solid food until they are a year old or older. The importance of supplemental foods and early weaning for human evolution are discussed in detail in chapter 12. Our ability to provide our babies with food other than mother's milk at an early age appears to have affected enzymatic and regulatory aspects of our milk composition. A reduction in the length of lactation, especially the length of exclusive breastfeeding, is an important element in our evolution from an ape-like existence to a human existence.

Protein-to-Energy and Relative Growth

It makes intuitive sense that the amount of protein an infant consumes should relate to its growth rate. Infants of species with fast growth rates should have greater protein requirements compared to infants with slower growth rates. This idea has been brought forward many times (e.g., Bernhart, 1961; Powers, 1933; Power et al., 2002) but never addressed very satisfactorily. There are a number of issues that complicate evaluating the extent to which growth rate and protein concentration in milk are linked. To compare growth rates of mammals that differ greatly in size is difficult. For example, in anthropoid primates adult size ranges from just over 100 g in the pygmy marmoset to 100 kg in a female gorilla, with male gorillas weighing up to 180 kg. A neonatal pygmy marmoset weighs about 20 g; a gorilla neonate about 2,000 g. A baby gorilla is 100 times bigger than a baby pygmy marmoset; but an adult gorilla is 1,000 times larger than an adult pygmy marmoset. The biology and allometry of growth is a complex subject.

Still, qualitatively most researchers agree that, as a broad generalization, New World monkeys grow relatively faster than do Old World monkeys,

Table 7.2. Anthropoid primate milk gross energy (GE) and protein content on an energy basis

Genus	Milk GE kcal/g	% of GE from protein	Protein per GE mg/kcal
New World monkeys			
Alouatta	0.56	23.3	40.0
Saimiri	0.91	23.3	39.7
Leontopithecus	0.64	22.3	38.1
Callithrix	0.76	20.8	35.5
Aotus	0.91	17.9	30.5
Cebus	0.89	15.9	27.2
Old World monkeys			
Macaca	0.90	12.7	21.6
Papio	0.82	10.7	18.3
Apes			
Gorilla	0.57	11.3	20.2
Pan	0.54	10.8	18.4
Pongo	0.54	8.6	14.7
Homo	0.68	8.6	14.7

Note: Milk protein content is highest in New World monkeys and lowest in the apes, especially low in milk of orangutans and humans. This order parallels the relative growth rates, with New World monkeys growing the fastest and orangutans and humans the slowest.

and apes are slower growing still. Within the anthropoid primates the New World monkeys generally have milks with the highest proportion of energy from protein, followed by the Old World monkeys and then the apes. Humans and orangutans appear to have the lowest proportion of milk energy from protein (table 7.2). The protein content expressed on a per energy basis matches the assessment of growth rate within primates; high protein-per-energy is associated with a relatively faster growth and development (Power et al., 2002; Power et al., 2008). This issue is discussed in more detail in chapter 12, where we consider the ways that human milk has evolved since the divergence of our lineage from that of the great apes.

Lactation Strategies in Summary

This chapter has provided a general overview of lactation strategies and a few examples. A lactation strategy is situated within a species' reproductive strategy, which is situated within its overall life history strategy. There are ecological and phylogenetic constraints on all aspects of life history, including a species' lactation strategy, and a species' lactation strategy potentially imposes constraints on other aspects of its life history. Many moving parts all come together to form an evolutionary successful whole.

Phylogeny appears to play a large role in setting a species' lactation strategy. Marmoset monkeys are the size of lab rats, but they produce monkey milk not rat milk. Giant pandas are members of the order Carnivora that have become herbivores, but they produce a carnivore-like milk. Elephants and rhinoceroses are similar in size and ecology but produce different milks that appear to reflect their phylogenetic origins.

PART III

More Than Food

In the ancestral mammalian lineage, milk was the earliest mechanism through which mothers signaled biochemically to their offspring. In the monotremes and marsupials it is still the primary mechanism. In the placental mammals the placenta now plays a major role, but milk remains important. We propose that milk and lactation have fundamentally affected mammalian developmental physiology, resulting in plastic (within reason) developmental patterns that rely on signals from the mother (Power and Schulkin, 2013).

In this section we explore milk as a regulatory mechanism; the many ways milk guides the growth and development of the neonate. There are molecules in milk that can affect all aspects of neonatal growth and development. In part II we discussed nutrients, and in particular how the relative proportions of protein to non-protein energy in milk relate to differences in growth. In this section we discuss the non-nutritive molecules in milk that affect everything from priming the neonatal immune system, to gut development, to helping to establish the gut microbiome.

Milk and the Immune System

Babies are vulnerable to diseases and parasites; more so than adults. This is true for all mammals. Baby mammals have immature immune systems. One function of milk is to protect infants from pathogens. There are a host of antimicrobial, anti-pathogenic, and immune-function molecules in milk.

The best-understood milk factors that prime and assist neonatal immunity are maternal immunoglobulins: IgA, IgG, and IgM. Milk is a means by which a mother can pass her disease history to her offspring; a Lamarckian inheritance of acquired characteristics. But there are many other immune-function molecules and even immune cells, such as macrophages, in milk.

The Microbiome

Each person has a unique genome that interacts with the environment to produce the individual characteristics of that person. But we are more than the results of our own genome. We have microbial communities (microbiomes) living on and inside our bodies that have coevolved with us. The amount of genetic information (DNA) represented by those microbial communities far exceeds our own genome. The gene products from those microbes contribute to our physiology and metabolism, in both positive and negative ways. The gut microbiome is perhaps the best known and studied, but there are microbiomes on our skin, in our eyes, in our mouths, and in the reproductive tract. There also appears to be a milk microbiome. Milk had been thought to be a largely sterile fluid; that was one of the putative advantages of breastfeeding. Now it is known that there are microbes in milk.

In addition, substances in milk act both as anti- and prebiotic factors. Many of the antibiotic molecules in milk are identical to immune factors that operate throughout the body (e.g., lysozyme c, pro- and anti-inflammatory cytokines, and so forth). However, there are milk-specific molecules, as well. For example, milk contains many oligosaccharides (long-chain, complex sugars) that are poorly if at all digested by mammalian digestive enzymes, and thus are not likely to be absorbed and metabolized by the infant. Instead, these complex sugars appear to have both prebiotic and antibiotic activity, encouraging the growth of commensal and symbiotic organisms and reducing the risk of infection by pathogens.

Milk Hormones

Modern research has found that milk may provide regulatory signaling that directs infant growth and development. In placental mammals, biochemical signaling from mother to offspring takes place throughout gestation through the maternal-placental-fetal connection. After birth, with the direct, physical connection between mother and infant severed, maternal biochemical signaling continues through substances she secretes into her milk. Gene products and other vital biochemical molecules necessary for growth, development, and survival are initially produced not by the neonate but by its mother and delivered to the neonate via milk. These information molecules of maternal origin have profound effects on infant development, from guts to brain. We review the growing body of evidence for maternally driven effects on infant development via milk.

Regulation of Gene Expression

One of the most exciting recent discoveries is that milk contains RNA molecules. Not surprising, of course, as any secretion is likely to contain biochemical materials from the secreting cell including RNA molecules. But in this case the RNA molecules are packaged in exosomes and microvesicles that appear to protect the RNA molecules from degradation by acids and enzymes, implying that these RNA molecules likely arrive at the neonate's small intestine intact. These RNA molecules include messenger RNA and what are termed interfering RNAs. The existence of these RNA molecules and their possible survival through the neonatal digestive process raise the possibility that neonatal gene expression may be directly affected by milk ingestion. In chapter 10 we review the evidence.

Developmental Origins of Health and Disease

The developmental origins of health and disease (DOHaD) represents a new, fundamental paradigm for investigating the metabolic diseases of modern humans (Barker, 1998; Gluckman and Hanson, 2004; 2006). In very basic terms, events during early life, including in utero, affect development. These developmental changes in early life carry over into adult life and contribute to the variation in health and disease risk among adult humans. Obesity, diabetes, and cardiovascular disease, among other human adult-onset diseases, all have potential origins in early life.

Although mothers of all taxa provide biochemical signals to their offspring, for non-mammals the time periods in which they can provide such signals are usually limited. For many taxa it is limited to the biochemical resources deposited into the egg. For placental mammals, such as humans, maternal biochemical signaling is a continuous process throughout the period of maternal dependence, starting with signaling through the maternal-placental-fetal axis and continuing after birth via milk. These signals are vital for the appropriate growth and development of the fetus and neonate.

Mammalian mothers can vary these signals, depending on circumstances. Certainly mammalian maternal signaling changes in a time-appropriate way. For example, in many mammals, milk composition varies considerably over lactation. This is especially true for marsupials; not surprising considering that marsupial neonates are extremely undeveloped. The appropriate milk to support growth and development of what is essentially a fetus is not likely to be the same as milk that supports growth and development of a joey that is almost ready to leave the pouch. But the signals may also vary due to maternal circumstances, such as her health or nutrition status. Now this has been shown to be true for other species, such as birds, as well. The biochemical composition of a bird egg will vary due to maternal condition at the time of egg formation. But after laying the egg the mother has little ability to change her biochemical signaling due to later environmental conditions. What is unique in mammals is that maternal circumstances may alter important signaling at multiple time points. Maternal circumstances at ovulation may have little to do with the path of development; the essential time periods during which development can be modified by maternal signals may occur at many time points between ovulation and weaning. The ability to modify offspring development due to environmental conditions over the extended period of maternal dependence is a fundamental mammalian trait. In essence, DOHaD appears to be an evolved adaptation in mammals, providing selective advantage in the broad perspective but sometimes leading to disease in the modern human world where we have changed some of our environmental inputs far beyond their evolutionary norms.

Chapter 8

Milk Protects

Life on Earth started billions of years ago as simple organisms without much structural organization. The advent of the cell was a major advance. Eventually multicellular organisms arose; but the single-cell organisms still are with us, and they affect our lives in both positive and negative ways. We have multiple microbiomes that coexist with us; in our guts, on our skin, in our eyes, indeed, in every cavity and compartment of our body there is a co-evolved microbiome. There are also many pathogenic microorganisms in the environment, toward which we have evolved multiple methods to avoid, kill, or at least survive. Babies are vulnerable to diseases and parasites; more so than adults. Adult organisms have survived the gauntlet and have proven defenses. Neonates are, for the most part, immunologically naïve and at exceptionally high risk from microbial pathogens, as well as depauperate in the composition of their own microbiomes. This is true for all mammals. Baby mammals have immature immune systems and immature microbiomes.

Mother's milk is among the first lines of defense against pathogens and among the first mechanisms to establish an appropriate, healthy microbiome in the neonate. There are a host of antimicrobial, anti-pathogenic, immune-function, and prebiotic molecules in milk. For example, milks contain many oligosaccharides that are poorly if at all digested, absorbed, and metabolized by the infant, but instead have both prebiotic and antibiotic activity. Milk also contains immunoglobulins, which means that milk is a mechanism by which a mother can pass her disease history to her offspring; an example of Lamarckian inheritance of acquired characteristics.

Antimicrobial and Immunological Aspects of Milk

Milk and its ancestral secretion probably always had functions related to the microbial world. For all the advantages that the original proto milk provided to eggs and/or hatchlings, what is good for eggs and neonate is also potential food for pathogenic microorganisms. From the beginning of proto lactation by our synapsid ancestors, mammary secretions would have contained antimicrobial factors for protection of both mother and offspring. At the same time, our gut microbiome is of considerable importance for our health and well-being. This was likely true for our synapsid ancestors of hundreds of millions of years ago. Mother's milk provides food not only for the neonate but also for the neonate's gut microbes. Milk probably has always contained prebiotic as well as antibiotic substances. Milk must serve multiple purposes; it has evolved to be inimical to some microbes and nurturing to others.

The immune system is traditionally divided into the innate immune system and the adaptive or acquired immune system. The innate immune system is the evolutionary older system and is found in plants as well as animals. The acquired immune system is found in all vertebrates. Both were active and functional in early synapsids. Milk likely has been associated with both immune systems for most of its evolution.

The innate immune system is comprised of a number of nonspecific defenses against potential microbial invaders. It is, in effect, a catchall category of defenses against infection that are not part of the adaptive immune system. Elements of the innate immune system include physical and chemical barriers, the inflammatory response, the complement system, and specialized phagocytic cells, such as natural killer cells. Milk does not provide a physical barrier but does contain elements from all of the other elements of the innate immune system.

The acquired immune system or adaptive immune system is composed of specialized cells and processes that confer immunity to specific pathogens based on exposure. Some of these cells (memory B and T cells) provide an immunological "memory" of an antigen associated with a previously encountered pathogen. The adaptive immune system is much more pathogen-specific, such that different individuals will have different adaptive immune systems while having essentially the same innate immune system. Milk contains maternal immunoglobulins that provide the immunologically naïve neonate with an initial adaptive immune response that is based on maternal pathogen exposure.

Thymus and Breast Milk

Interestingly, the thymus gland of breastfed infants is significantly larger than that of formula-fed infants. The thymus is an ancient gland (it is found in all jawed vertebrates) that is critical to the production of T cells. Thymus size is associated with health and well-being of infants. A small thymus in early infancy is a risk factor for all cause- and infection-related mortality in the first years of life in Guinea-Bissau (Aaby et al., 2002; Garly et al., 2008) and Bangladesh (Moore et al., 2014). In the Bangladesh study, thymus size at birth was not associated with mortality, while thymus size at eight weeks of age was, suggesting that early postnatal influences, possibly including factors in breast milk, affect thymus development during this time period. In rural Gambia, infants born during the "hungry" season have smaller thymus glands (Collinson et al., 2003) and shorter life spans (Moore et al., 1999) than infants born during the harvest season, with a peak difference in thymus size at eight weeks. At eight weeks of age infants born in the hungry season had a correspondingly lower thymic function (Ngom et al., 2004). At one and eight weeks of infant age breast milk collected during the hungry season had lower concentration of IL-7 compared to breast milk collected during the harvest season at the same infant age (Ngom et al., 2004), suggesting a possible proximate link between the transfer of IL-7 from mothers to their babies via milk and neonatal thymus gland development.

The Milk Immune System

For most of the human history of the study of milk the focus was on nutrition; milk as food. It has only been within the last 60–65 years that the immune components of milk have been studied. That may seem like a long time (the first paper on antibodies in milk was prior to either author's birth), especially considering recent discoveries of factors in milk such as micro RNAs (within the last 5–10 years), but milk as a food has been studied for much longer. The initial immunological work was difficult and the evidence often indirect due to the lack of technology to detect the immune components in milk. It wasn't until the 1970s that the concept of an immune system in milk was truly considered (Goldman and Smith, 1973).

There was evidence for an immune function of milk from much earlier. Paul Ehrlich reported evidence for the transfer of immunity to specific pathogens by murine milk in the 1890s (Ehrlich and Hubner, 1892). Jules Bordet demonstrated bacteriolytic properties of human milk in 1924 (Bordet and Mordet, 1924). However, it wasn't until more than 40 years later that the active agent, lysozyme c, was identified (Chipman and Sharon, 1969). Lactoferrin in milk was discovered in 1961 (Blanc and Isliker, 1961) and its antibacterial properties soon after (Masson et al., 1969). The existence of immunoglobulins in human milk, especially colostrum, with a preponderance of a form of IgA, was verified about this same time (Hanson, 1961).

There was also an expanding amount of epidemiological evidence collected since the 1950s that demonstrated a greater resistance to many diseases by breastfed infants relative to infants fed cow's milk or other artificial formula (Mata and Wyatt, 1971). The association between weaning and diarrheal disease was noted (Gordon et al., 1963). Breast milk was known to have beneficial effects including lowering the risk of diarrhea, ear infections, and respiratory infections. The mechanisms were poorly if at all understood, until the technology improved to be able to detect more of the immune factors in milk. An important discovery was that human milk contained living immune cells; neutrophils, macrophages and lymphocytes (Smith and Goldman, 1968). This discovery indicated a functioning immune system within milk, not just immune function molecules secreted into milk (Goldman, 2007).

There are a number of important questions regarding the milk immune system to which science is beginning to provide answers:

1. For any immune factor, does it provide protection for the mammary gland, the neonate, or both?
2. To what extent do these immune factors accommodate a lack in the neonate versus supplementing factors already expressed by the neonate?
3. Is the main function protection against infectious disease? Or do these milk-borne factors have developmental, pro- and/or anti-inflammatory, or other actions that can affect health beyond infection?
4. To what extent does breast milk interact with the maternal and neonatal microbiomes?

The last two questions are particularly important, in our opinion, and raise a complication for the organization of any book on the biological functions of milk. This chapter addresses milk's immune function, and these questions are appropriate to be addressed here. But questions 3 and 4 are also appropriate for developmental and regulatory effects of milk, which we address in the next two chapters. The immune system and the developmental/regulatory system are not separate entities. They have overlapping functions and effects, and all act within the same organism. Evolved systems frequently have overlapping functions for different molecules as well as multiple functions for each molecule. Science is demonstrating that our microbiomes have substantial effects on all aspects of our biology. Lactation and milk are ancient, coevolved systems with actions on mother, child, and microbiomes. The immunological and developmental/regulatory functions of milk do not in reality easily separate into two entities. The next three chapters are necessarily intertwined.

Immunoglobulins

For most mammals immunoglobulins do not pass the placenta; so milk is the first and only way to transfer maternal acquired immunity. Anthropoid primates are different in that IgG can pass the anthropoid hemochorial placenta (Coe et al., 1994). Other placental types (epithelialchorial and endothelialchorial placentas) apparently cannot pass immunoglobulins. In species with those placenta types (e.g., horses, dogs, prosimian primates) little or no maternal immunoglobulin is transferred during gestation, and milk provides the only avenue for transmission of maternal immunity to offspring. In the milks of these species there are significant concentrations of IgG and IgM as well as IgA (Van de Perre, 2003). Regardless, mothers are the source of infant immunity to a variety of diseases, based on the mother's own disease history.

Milk immunoglobulin content has been well studied in humans and domesticated mammals such as the cow, sheep, horse, and pig. For many ungulate species, most immunoglobulin transfer occurs in the immediate few days postpartum. The neonatal intestinal tract will pass macromolecules in the immediate postpartum period. Intestinal permeability decreases, often rapidly, but somewhat differently across species and, for some species, can be affected by milk intake.

The primary milk immunoglobulin in anthropoid primate milks is secretory IgA (SIgA), whose main site of function is mucosal surfaces, including those of the gut (Corthésy, 2007). Maternal SIgA transferred to the neonatal intestine via milk is capable of binding to a wide variety of pathogenic microorganisms in the lumen, preventing their adhesion to mucosal surfaces and thus their entry into epithelium (Corthésy, 2007). SIgA also is able to adhere to special cells (M cells) in what are called Peyer's patches distributed primarily in the lower section of the small intestine. Binding to M cells allows the uptake of SIgA and SIgA-coated bacteria across the intestinal epithelium where it can interact with dendritic cells. The immunoprotective action of milk-borne SIgA can last throughout lactation, well past the time when other immunoglobulins would be unable to pass the intestinal mucosa. Significant levels of SIgA are detectable in human, gorilla, and orangutan milk from colostrum probably through the end of lactation. We have unpublished data showing detectable SIgA in gorilla and orangutan milk up to three years after the birth of the neonate.

In addition to its role in blocking adhesion and entry to epithelium by pathogenic microbes, SIgA appears to have a role in programming immune tolerance to commensal microbes and potential antigenic proteins (Corthésy, 2007; 2013). SIgA also has non-inflammatory properties as it does not trigger the release of inflammatory mediators. Thus, SIgA plays a major role in maintaining intestinal epithelial homeostasis, protecting it from infection, moderating immune responses to beneficial microbes and potential antigenic peptides, while minimizing inflammatory responses (Corthésy, 2013).

Cells

Milk also contains maternal leukocytes, which function both to protect the mammary gland from infection (e.g., neutrophils and macrophages) and to assist the neonatal immune system (e.g., macrophages and lymphocytes). Macrophages in breast milk are phagocytic, and thus capable of attacking pathogens in milk and in the neonatal intestinal tract, but they also express T and B cell activation markers. The majority of milk lymphocytes are T cells. Maternal macrophages and activated T cells not only can help compensate for the immature neonatal immune system, but also may promote its maturation (Field, 2005). Many of the cytokines and other peptides found in milk may derive from the immune cells in the milk rather than or in addition to being secreted by mammary epithelial cells. Thus, the cellular fraction of milk may also play a major role in milk's potential endocrine and cytokine function, discussed in the next chapter.

Peptides

There are a variety of peptides found in milk that have antimicrobial function. They may have other functions as well. Some may end up digested by the neonate and provide amino acids, but their main function would appear to be protecting the mammary gland and/or the neonatal gut from microbial pathogens. Two of the first to be discovered were lysozyme c and lactoferrin, which we have discussed previously. Others include lactadherin, lactoperoxidase, and mucins. Some are found in diverse tissues and organs throughout the body (e.g., lysozyme c) while others appear to have a more restricted distribution. Only α-lactalbumin is restricted to milk, and its antimicrobial properties are somewhat uncertain in real-world settings.

Lactadherin, now known as milk fat globule-EGF factor 8 protein (MFG-E8), was first identified as a major protein component of the mouse milk-fat globule (Stubbs et al., 1990). It has been shown to be expressed by mammary tissue from many mammal species, but also to be ubiquitously expressed in many organ tissues besides the mammary, with high expression in lung, spleen, lymph nodes, and brain and low expression in intestine and liver (Matsuda et al., 2011). Current research focuses on its role as a "bridging molecule" that enhances the phagocytic removal of apoptotic cells. Low levels of MFG-E8 are associated with increased inflammatory response due, in part, to lower clearance of apoptotic cells (Matsuda et al., 2011).

Lactadherin may have a different function in milk. In human milk, a form of lactadherin glycosylated with a sialic-acid-containing oligosaccharide binds to rotovirus. Mothers' of asymptomatic infants with rotoviral infection were more likely to have above-average concentrations of lactadherin in their milk (Newburg et al., 1998). Neither the non-glycosylated form of lactadherin nor free human milk oligosaccharides display significant rotovirus-binding function, implying that it is necessary for the lactadherin molecule to be glycosylated with a sialic-acid-containing milk oligosaccharide in order to have anti-rotoviral activity.

Many proteins in milk serve multiple functions. For instance, mucins are found in the milk fat globule and are important components of the membrane that encapsulates milk fat. Mucins also have antimicrobial action. At least two of the 16 mucins have been identified in milk: mucin 1 and mucin 4. One anti-pathogen mechanism of action for mucin 1 is pathogen binding to a sialic acid residue which then inhibits binding to epithelial cell surface glycans (Liu and Newburg, 2013). Mucins 1 and 4 have been shown to inhibit infection by *Salmonella* and mucin 1 has been shown to interfere with rotavirus, *E. coli*, and HIV transmission (Liu and Newburg, 2013).

Lactoperoxidase is secreted by mammary, salivary, and other mucosal glands. It has antimicrobial properties. Lactoperoxidase is found in high concentrations in bovine milk, but only low concentrations in human milk. It is found in high concentrations in human neonatal saliva, however. This illustrates an important concept regarding many bioactive molecules in milk. They usually are molecules that the neonate will eventually begin to produce for itself. Milk provides a transition between the uterine environment in which many resources were wholly provided to the infant by the mother and the world after birth where the direct connection with the mother no longer exists. In some cases milk is, for a while, the sole provider of a crucial biochemical; in others, milk supplements neonatal production as the neonatal systems develop and mature. Species differ in the molecules neonates rely on obtaining from milk. For a bovine calf, mother's milk is the source of lactoperoxidase in the intestinal tract to combat pathogenic microbes; for a human baby, its own saliva is the source.

Oligosaccharides

The main oligosaccharide in milk is the disaccharide lactose. Lactose has important function in providing energy to neonates and has some direct regulatory effects on intestinal epithelium, but may not be particularly important for the purposes of this chapter. It has long been known that milk, especially human milk, contains sugars in addition to lactose. In this chapter we consider the prebiotic and antibiotic functions of larger milk oligosaccharides. These milk oligosaccharides are largely indigestible and will travel through the gut down to the large intestine. Some of these oligosaccharides will be digested and metabolized by microbes in the colon. Interestingly, the microbes that preferentially utilize these milk oligosaccharides as a food source are the co-evolved, generally beneficial microbes found in a healthy human gut (Ward et al., 2006; Sela and Mills, 2010). They protect the neonate against pathogenic microorganisms by out-competing the pathogens, but they also may be important for the appropriate development and maturation of the neonate's immune system.

Many milk oligosaccharides are resistant to microbial digestion. These glycans appear to act as decoy attachment molecules. Epithelial cell surfaces are coated with oligosaccharides, which serve as attachment sites for molecules important in cell-to-cell communication. These cell-surface oligosaccharides also serve as attachment sites for microbes, providing access to the cell. Pathogens that bind to a soluble milk oligosaccharide in the neonatal gut instead of to its target cell-surface oligosaccharide will be incapable of infecting the neonate and end up being safely defecated.

In a study using a cell line derived from intestinal epithelial cells from premature infants, human milk oligosaccharides were shown to be effective in reducing invasion of the cells by the fungus *Candida albicans* (Gonia et al., 2015). This fungus causes a majority of the intestinal fungal disease in premature infants. The mechanism for the antifungal properties of the milk oligosaccharides is unclear.

Necrotizing Enterocolitis

The most common gastrointestinal pathology seen in neonatal intensive care units, necrotizing enterocolitis (NEC) is characterized by hemorrhagic inflammatory necrosis of the intestine, primarily the distal small intestine and proximal colon, and is a major source of morbidity and mortality in preterm babies (Sullivan et al., 2010). The etiology of this disease is uncertain. The immaturity of the preterm infant intestinal tract may cause it to be more susceptible to inflammatory disease. Some researchers have suggested that NEC derives from interactions with the developing microbiome, due to the immature immune system or to an inappropriate microbiome composition. Treatment often requires surgical removal of affected tissue.

Feeding preterm infants human breast milk reduces their risk of developing NEC. The specific signaling molecules in breast milk that act on the infant's intestinal tract are not known, although EGF and TGF-β-2 are logical candidates. Both of these growth factors have significant effects on intestinal epithelial cell proliferation and maturation (Coursodon and Dvorak, 2012). Both of these molecules are found in amniotic fluid (Underwood et al., 2005), which the preterm infant would still be swallowing if it had not been born. A reasonable hypothesis regarding the etiology of this debilitating disease is the lack of an exogenous signal provided by the mother, first via amniotic fluid and after birth via milk, which results in pathologic development of intestinal epithelial cells in the neonate and a dysfunctional immune response with excessive inflammation.

Other researchers have suggested that an inadequate/inappropriate intestinal microbiome may play a role in NEC (Schwartz et al., 2012). NEC is intrinsically an inflammatory disease, implying that interactions between the gut immune system and its microbiome could have potent effects on vulnerability or resistance. Interactions between the neonatal gut immune system and the developing microbiome are thought to be important for the appropriate maturation of the immune system. The neonatal microbiome that co-evolved with us will have a low inflammatory potential; a gut microbiology ecology that deviates from the co-evolved one may present a greater inflammatory challenge.

Intestinal epithelial cells are constantly being sloughed (exfoliated) into the intestinal lumen (Schwartz et al., 2012). These cells provide an opportunity to examine gene expression by gut epithelial cells in a non-invasive manner, as they can be extracted from feces. Breastfed babies develop a different intestinal microbiome compared to formula-fed babies (Bezirtzoglou et al., 2011). They also display different patterns of gene expression from exfoliated epithelial cells in the feces (Donovan et al., 2014). The gene expression pattern in intestinal epithelial cells of breastfed babies indicates a lower inflammatory state and greater resistance to the effects of hypoxia, consistent with their greater resistance to NEC.

The Interconnected Microbiomes

One of the more fascinating and potentially productive discoveries of modern biology regards the extent and importance of the co-evolved microbial communities that live on us. Microbes live on and in virtually every part of our body: skin, eyes, ears, nose, mouth, gut, lungs, and probably every other organ or orifice that has contact with the outside environment. It has been well known that microbes are ubiquitous, but until recently their detection has relied on culture methods that simply do not work to detect the majority of co-evolved microbes inhabiting niches on and in the human body. The advent of next-gen sequencing techniques and the production of libraries for the 16S ribosomal RNA gene (box 8.1) has resulted in the detection of microbial communities in many body fluids/regions previously thought to be relatively sterile, including placenta, amniotic fluid, and milk. Major research efforts are beginning to produce an understanding of the diversity and extent of the microbial communities that have co-evolved with us (e.g., International Human Microbiome Consortium; NIH Human Microbiome Project).

The total mass of microbes on and in our bodies is about 3% of our total weight, and it represents hundreds of times more DNA (and genes) than is in our own genome. Our gut alone contains at least two orders of magnitude more expressed bacterial DNA than the entire human host expressed genome. Gut microbes produce a large array of biologically active products that affect other gut residents, gut epithelium, and even are absorbed into circulation. These microbial gene products interact with our own expressed genome to regulate and influence our metabolism, physiology, and immune function, and are very important for health and resistance to disease.

Microbiomes are microbial communities that form microbial ecologies. Microbes produce products both necessary for and inimical to other microbes resulting in biochemical cooperation/commensalism as well as biochemical warfare. The microbial community sustains and nurtures itself

---

BOX 8.1

16s ribosomal RNA

Carl Woese and George Fox pioneered the use of ribosomal RNA (rRNA) to create consistent phylogenies of single-celled organisms (Woese and Fox, 1977).
Ribosomal RNA is the structural RNA component of the ribosomes. Ribosomes
are found within all cells and are the site of protein synthesis. Ribosomal RNA
form complexes with proteins to create two subunits, called the large and small
subunits. In prokaryotes, the small subunit is termed the 30S rRNA; it contains
the 16s rRNA. The 16S rRNA has important structural and binding functions necessary for ribosomes to perform their protein synthesis function.

The genes coding for 16S rRNA are strongly conserved within phylogenies,
with slow rates of evolution. This makes them ideal to sequence for the purpose
of constructing prokaryote phylogenies, or to determine which prokaryote taxa
are in a sample. The use of 16S rRNA DNA sequencing has been vital to detecting
the large number of microbial taxa in the gut (and other microbiome locations)
that cannot be cultured. Many microbes require the unique, coevolved environment of their microbiome in order to survive and replicate. Removing them from
their habitat kills them. Before this technique of sequencing 16S rRNA DNA, if a
microorganism could not be cultured, it was virtually invisible. But regardless of
their ability to live outside of the gut, the DNA of all gut microbes can be extracted and sequenced. Our window into the world of the gut microbiome has
become larger and clearer.

---

but resists invasion by outsider microbes. A well-established, co-evolved
microbiome benefits the host by this resistance to potential pathogens.
The definition of the innate immune system could be stretched to include
healthy microbiomes on skin, gut, and so forth, providing a barrier against
infection.

The focus of this section is on the milk microbiome, but the different
microbiomes are interconnected in many ways. For example, the oral cavity microbiome shows a strong association with the newly discovered placental microbiome (Aagaard et al., 2014). Gut and vaginal microbes are
also found in the placenta. Most directly relevant for this book, the microbial communities in mother's milk appear to be influenced by the maternal
gut microbiome (Martín et al., 2012). And of course the infant's gut and
oral cavity microbiomes will likely influence and be influenced by the milk
microbiome.

Gut Microbiome

It has been long known that different portions of the digestive tract can harbor large microbial communities. Microbes play a substantial role in the digestion of food, especially fibrous foods, for many animals. Whole chambers of the GI tract have evolved to house symbiotic microbial communities that assist in digestion, provide important nutrients, and affect metabolism. From cows, with their complex, four-chambered stomachs filled with fiber-digesting bacteria, to rabbits, where the fermenting organisms primarily live in the cecum and colon, microbes contribute to the nutritional well-being of animals.

In the 1880s, Dr. Theodor Escherich, arguably the first pediatric infectious disease physician (Shulman et al., 2007), isolated 19 different bacteria from the feces of neonates. He demonstrated that meconium was sterile (we now know that is actually not true), but that bacterial colonization of the neonate's gut occurs within the first 24 hours after birth. He described in detail the common colon bacillus which he called *Bacterium coli commune*, but is now named after him: *Escherichia coli*. In 1899, Henry Tissier, a French pediatrician at the Pasteur Institute in Paris, isolated a bacterium characterized by a Y-shaped morphology ("bifid") in the intestinal flora of breastfed infants and named it "bifidus." Tissier and Theodor Escherich's student, Ernst Moro, demonstrated that breastfed infants had different microbial communities in their feces than did non-breastfed babies. In addition, compared to the feces from healthy babies, babies with diarrhea had fewer "bifidus" bacteria in their feces.

The gut microbiome produces a large array of biologically active products that can affect other gut residents, gut epithelium, and even be absorbed into circulation. An aberrant gut biome is associated with increased disease risk in infants (Isolauri, 2012). Interestingly, obese women produce milk with a less diverse set of microbes, and women who gain excessive weight during pregnancy produce milk with different microbes from those in milk of women who had normal weight gain (Cabrera-Rubio et al., 2012). Obesity in adults is associated with an altered gut microbial population.

The establishment of the gut microbiome is a vital aspect of infant health and well-being and is hypothesized to have direct and indirect effects on growth, development, and the vulnerability to obesity and obesity-related diseases. Recent studies have shown the importance of the gut microbiome in human physiology and metabolism. For example, malnutrition has been linked to differences in the gut biome in a study of twins in Malawi. There was evidence that the gut biome differences not only potentially affected digestive and absorptive efficiency, but were actually related to changes in regulation of aspects of the Krebs cycle in these children (Smith et al., 2013).

Milk Microbiome

Milk was previously considered to be a relatively sterile body fluid, which was one of the putative advantages to breastfeeding. Microbes found in milk were thought to represent contamination from skin or the environment, or a sign of infection in the mammary gland. However, next-gen sequencing techniques and the development of libraries for the 16S ribosomal RNA gene have shown that milk contains a microbiome. Of significant interest is the detection of symbiotic gut microbes in breast milk. Thus, milk may represent one mechanism for the inoculation of the infant gut with maternal gut microbes, presumably aiding in establishing a healthy infant gut microbiome from an early age.

The stability of the milk microbiome has not been well studied. Longitudinal samples of milk are required to test whether the microbiome changes over lactation. Evidence from human studies and our own data on the gorilla milk microbiome indicate that there may be consistent changes in milk microbiome composition over lactation as the infant ages. More research is needed to examine what, if any, significance these changes might have.

Mouth Microbiome

Mouths are dirty places, full of microbes. Not surprising considering that mouths are kept moist and warm and foreign material is constantly being placed inside them. To eat, drink, and even breathe is to take microbes into the mouth. And, of course, residual food that remains in the mouth provides a banquet for any microbes residing there. The mouth is a great microbial habitat. Knowledge about microbes in our mouths dates back to the earliest instances of observations of microbes. Antonie van Leeuwenhoek was the first to see one-celled organisms (he called them animalcules) in a variety of fluids, including from his and other people's mouths. So microbes in our mouths have been known since around 1676.

Different tissues within the oral cavity appear to support different microbial communities. The microbes on the tongue, teeth, below the gum line, palate, tonsils, and so forth differ in community composition. The oral cavity microbiome rivals the gut microbiome in complexity and diversity and numbers of taxa (Human Microbiome Project Consortium, 2012).

Microbes in the mouth can easily transfer into the bloodstream. This is thought to be the mechanism that links the oral microbiome to the placental microbiome, and a potential proximate mechanism for why periodontal disease may be related to an increased risk for preterm birth.

MOUTH TO MAMMARY AND BACK

Microbes in the infant's mouth could come from milk, be influenced by milk (e.g., by milk nutrients or immune function molecules), and could travel from the infant's mouth to the nipple and into the mammary gland. Imaging studies of the infant's oral cavity and the breast during suckling have shown that there is retrograde flow; fluid goes from the baby to the mammary as well as from the mammary to the baby. Thus, the baby's mouth may be a source of microbes that will be found in milk and vice versa. This might be one mechanism to explain the changes in milk microbiome over lactation mentioned above.

Preparing the Infant Gut Microbiome

Breast milk is a major avenue and facilitator for infant gut microbe colonization. Milk contains prebiotic factors, primarily the milk oligosaccharides already discussed, that feed symbiotic and commensal microbes that take residence in the neonatal gut. Evidence from human and rodent studies suggests that maternal breast milk might provide a crucial source of microbes that initially colonize the infant gut (Martín et al., 2004; Fernández et al., 2013). Transfer of live gut bacteria to mesenteric lymph nodes and to the mammary gland occurs during late pregnancy and lactation in mice (Perez et al., 2007). Maternal characteristics affect the milk microbiome (Cabrera-Rubio et al., 2012; Collado et al., 2012). Obese women produce milk with a less-diverse set of microbes, and women who gain excessive weight produce milk with different microbes from those in milk of women who had normal weight gain during pregnancy (Cabrera-Rubio et al., 2012). This suggests that obese mothers may pass on their obesogenic gut microbiome to their breastfed infants at least in part through their milk.

The gut microbial populations of breastfed infants differ from that of bottle-fed infants (Bezirtzoglou et al., 2011), and generally reflect their mother's gut biome. Breast milk contains staphylococci, streptococci, bifidobacteria, and lactic acid bacteria, and the same strains that are found in a mother's milk are usually found in her infant's feces (Martín et al., 2012). Many of the bacteria found in breast milk are also found in the maternal gut, and are considered to be commensal organisms (Martín et al., 2004). This suggests that breast milk is a vehicle by which mothers colonize their babies' guts with maternal gut microbes.

The origin of what appear to be maternal gut microbes in breast milk is somewhat controversial. The standard theory was that bacteria in milk were the result of contamination from maternal skin or the infant's oral cavity. And both these surfaces likely do contribute some microbes to the breast and into the milk. Recently, the existence of an enteric-mammary transport system that encapsulates bacteria from the maternal gut and transports it to the mammary gland has been suggested (Martín et al., 2012; Fernández et al., 2013). Dendritic cells have been shown to be able to penetrate the gut epithelium and take up bacteria from the lumen (Rescigno et al., 2001). Intestinal dendritic cells can retain small numbers of live bacteria for several days (Macpherson and Uhr, 2004). Macrophages may also be able to take up, retain, and then disperse live bacteria. Transfer of live gut bacteria to mesenteric lymph nodes and to the mammary gland occurs during late pregnancy and lactation in mice (Perez et al., 2007). The feces of infants born by Cesarean section to women given oral *lactobacilli* during gestation contain that strain of *lactobacilli*, even though they were not exposed to the vaginal environment (Schultz et al., 2004). Milk is the likely colonizing source, implying that live *lactobacilli* in the maternal gut can reach the mammary glands, enter the milk, and colonize the neonate. Thus, it appears that mothers can inoculate their infants with maternal gut microbes via milk, with potential physiological consequences for the infants.

Many microbiologists are skeptical that milk microbes play much of a role in inoculating the infant and starting the neonatal gut microbiome. They point to the many other sources of maternal microbes, including the birth canal. It is true that babies delivered by Cesarean section generally start life with a different gut microbial community (Grölund et al., 1999). Microbiologists also refer to the "fecal veneer" that exists, even in the most hygienic homes. It would be difficult for a baby to avoid exposure to its mother's gut microbes. These scientists stress the prebiotic factors in milk to explain the differences in gut microbiome between breastfed and formula-fed babies.

The transport system from maternal gut to mammary gland could serve an alternative function besides providing potential colonizing microbes for the infant. It could serve as a signal to the mammary gland to produce particular molecules or other responses cued by the microbes delivered. Perhaps to produce antimicrobials if the microbes transported to the mammary are potential pathogens. Perhaps to modify the type or concentration of prebiotic oligosaccharides depending upon what species of symbiotic microbes are transported. Milk oligosaccharide content, both quantity and types, vary tremendously in human milk. We are not aware of any definitive research to determine the extent to which variation in milk oligosaccharide composition can be induced, or whether the differences in milk oligosaccharides in milk samples from different women truly represent fixed differences between women. Is the milk oligosaccharide content constant over lactation? Is it consistent for a woman or can it vary between lactation events for the same woman? Can the mother's gut microbiome influence her milk oligosaccharide content? Important questions to be answered.

Chapter 9

Milk Guides

The neonatal and infant period is one of substantial growth and development. Milk certainly provides the nutritional building blocks for this growth and development. Milk may provide much of the regulatory signaling that directs infant growth and development as well. In placental mammals, biochemical signaling from mother to offspring takes place throughout gestation through the maternal-placental-fetal connection. After birth, with the direct, physical connection between mother and infant severed, maternal biochemical signaling continues via substances she secretes into her milk. Important aspects of early postnatal development appear to be regulated to a significant extent via milk. Breast milk contains a wide array of bioactive molecules that likely play a role in shaping growth and metabolism in the neonate (Savino et al., 2011; Bronsky et al., 2011). These include growth factors (e.g., the insulin-like growth factors [IGFs], epidermal growth factor [EGF], and transforming growth factor [β (TGF-β)]), metabolic hormones (e.g., ghrelin, leptin, and adiponectin), and cytokines (e.g., interleukins such as IL-6 and IL-8) that can directly impact infant physiology, particularly the developing gastrointestinal tract and immune system (Rautava and Walker, 2009). A broad spectrum of evidence suggests that these signaling molecules have important effects on growth and development of neonates. Studies using *in vitro* and animal model approaches show that milk bioactive factors such as IGF-I, TGF-ß, EGF, leptin, ghrelin, and adiponectin are involved in epithelial proliferation, gut differentiation, and suppression of gut inflammation, and shape growth and metabolism in the neonate (Wagner, 2002).

Breast milk is protective against childhood obesity (Savino et al., 2009) and may play an important role in the development of glucose regulatory physiology probably via milk-borne hormones such as leptin, ghrelin and adiponectin (Savino et al., 2011). Recent evidence suggests that mothers may pass on their metabolic history to their infants. Maternal obesity predisposes offspring to obesity and metabolic dysfunction in rodent models.

Cross fostering mouse pups from lean dams to obese dams during lactation results in those pups becoming vulnerable to obesity and metabolic disease (Gorski et al., 2006; Oben et al., 2010). The reverse, cross-fostering mouse pups that are predisposed to obesity onto lean dams lowers the extent of obesity in the pups (Reifsnyder et al., 2000) and improves insulin sensitivity (Gorski et al., 2006). Obesity-prone and obesity-resistant rat pups displayed a phenotype consistent with the foster (nursing) dam's genotype prior to weaning rather than to their own genotype (figure 9.1), although after weaning their food intake and weight gain reflected their own genotype (Schroeder et al., 2010). There was an effect of sex; male obesity-prone pups fostered to obesity-resistant dams, although still becoming obese post weaning, had significantly lower body weight and fat mass than obesity-prone male pups nursed by obesity-prone dams (figure 9.1; Schroeder et al., 2010).

Human studies to date have shown limited evidence for milk bioactive factors affecting growth and adiposity. Over the first six months of lactation, milk adiponectin concentration was shown to be negatively associated with weight-for-length and weight-for-age, but not with length-for-age, implying lower adiposity in infants drinking breast milk with high adiponectin (Woo et al., 2009). However, those infants experienced catch-up growth in the second year of life, with an increase in adiposity (Woo et al., 2012). Higher concentration of IL-6 in breast milk was associated with lower weight gain and fat mass in infants (Fields and Demerath, 2012). Small for gestational age (SGA) infants who are breastfed gain less fat mass than formula-fed SGA infants and maintain a healthier metabolic profile in terms of circulating IGF-1 and high-molecular-weight adiponectin (de Zegher et al., 2012; 2013). Previous studies using enriched formula have shown that SGA infants with greater catch-up growth end up with higher fat mass (Singhal et al., 2010) and higher blood pressure at 6 to 8 years of age (Singhal et al., 2007). The increased risk of cardiovascular disease in later life for SGA infants may in part be due to well-intentioned but possibly biologically unsound efforts to enhance weight gain. Whether the slower growth and better metabolic profiles of breastfed SGA infants is due solely to nutritional factors or whether signaling molecules in breast milk contribute is unknown. What has been shown is that the composition of milk fed to neonates influences their growth rate, body composition, and eventual metabolism, and that mothers' milk is more likely to produce a healthy outcome.

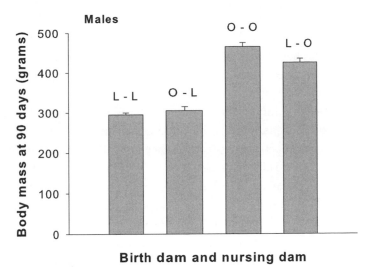

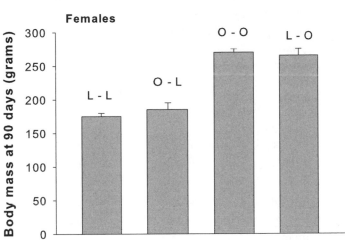

Figure 9.1. Body mass at 90 days for rat pups born to either a lean genotype dam (L) or an obese phenotype dam (O) and cross fostered to either a lean or obese genotype dam: L - L = lean-lean; O - L = obese-lean; O - O = obese-obese; L - O = lean-obese. Pup phenotype matched the nursing dam genotype, although O - O males were significantly larger than O - L males. Data from Schroeder et al., 2010.

Milk Oligosaccharides (Again)

Oligosaccharides have functions in the body completely outside of the pre-biotic and anti-pathogenic actions ascribed to the milk oligosaccharides. This raises a few interesting questions, without answers as yet, as far as we can discern. Does the importance of oligosaccharides in cell-to-cell commu-nication extend to signaling by milk oligosaccharides? Can milk oligosac-charides produce an activating signal if they bind to a cell membrane? Do any milk oligosaccharides become incorporated into neonatal tissue? Al-though the general statement is that milk oligosaccharides are not digested, absorbed, and metabolized, like many aspects of biology this does not appear to be 100% true. Largely indigestible does not mean completely indi-gestible. About 1% of human milk oligosaccharides are absorbed by the gut and enter circulation, as indicated by detection in neonatal urine (Bode, 2012). Modified milk oligosaccharide metabolites were found in the urine of breastfed infants, implying either the infant or the infant's gut microbiome had metabolized milk oligosaccharides which had then been absorbed into circulation (Dotz et al., 2015). Whether these absorbed milk oligosaccha-rides are capable of biological activity or are simply filtered out by the kid-neys and excreted is unknown.

Sialic acid is an important brain constituent. Many milk oligosaccha-rides contain sialic acid. Milk oligosaccharides are thought to be important in brain development for carnivore species that are born extremely altricial (bears being the prime example); carnivores have low-to-moderate sugar content to their milk, but relatively little is lactose with most being other oligosaccharides. Does the small amount of milk oligosaccharide that is ab-sorbed by a human neonate contribute at all to brain development?

Milk Hormones

Milk contains steroid and peptide hormones. These hormones undoubtedly exert physiological effects on infants, though the direct evidence is scant largely due to a paucity of research in this area, and, to be fair, the difficulty of doing intervention studies. Leptin in milk has been suggested to influ-ence neonatal milk intake and perhaps to influence later-life appetite and feeding behavior. We previously briefly described some findings regarding milk adiponectin and infant growth.

Cortisol concentration in milk has been shown to be associated with temperament in both human (Grey et al., 2012) and rhesus monkey infants

(Hinde and Capitanio, 2010; Hinde et al., 2015). Interestingly, milk energy transfer at 1 month of age also was associated with infant temperament at 3–4 months of age in rhesus monkeys (Hinde and Capitanio, 2010). Infants of mothers who produced above-average available milk energy (the product of milk energy content and milk production estimated by the yield of milk from complete evacuation of the mammary gland after separation from the infant for 3 hours) were more active and displayed a more confident temperament when separated from their mothers 3 months later. The interactions and associations between nutrients and hormones in milk have been little studied, but may be a fruitful area of future research. This leads us to the next section.

Geeky Technical Note on Measuring Hormones in Milk

We apologize to the reader; we know we said in our introduction we would put items like this in a box which then could be read or avoided. But we think this is an important technical aspect of the modern investigation of milk, specifically regarding measuring hormones in milk, which has been overlooked to the possible detriment of the field. Put simply, most published papers reporting the hormone content of milk simply report the concentration (e.g., ng/ml) without any adjustments for the possible variations in milk composition that occur between women and even in samples of milk from the same woman collected at different times or under different protocols. This variation often co-varies among multiple milk constituents, and possibly arises from similar causes. For example, the rate of lactose synthesis not only determines the lactose content of the milk but also affects the osmotic gradient that draws water into the mammary gland and dilutes all milk constituents. Some of the measured variation in milk constituents may be largely explained by variation in water content, with little biological significance.

In urine, hormone levels are rarely, if ever, expressed as a straight concentration as measured. Usually, hormone concentration is adjusted to creatinine, in part to account for the large potential variation in water content of urine. Even hormone concentrations in blood samples are affected by fluid volume. For example, pregnant women show a decrease in circulating hormone levels that is related more to increased blood volume (more water) than to any real changes in hormone production/secretion. Milk is a regulated fluid; but it does not appear to be as tightly regulated as blood, though it seems to be more regulated than urine. Rather than comparing absolute amounts, the relative proportions of milk constituents, including hormones, may better reflect the biological significance of differences between samples, individuals, and species.

The total amount of hormone (or any other bioactive factor, such as a nutrient or immunoglobulin) transferred per day or per nursing out via milk depends upon the concentration in milk, but also on the volume of milk consumed. Although the factors that regulate milk intake by infants have not been fully determined, it is reasonable to propose that, in early infancy, milk intake must satisfy energy requirements. Infants consuming a low-energy milk are likely to ingest a larger volume of milk to meet their basic needs. Comparing concentrations of any milk constituent between different species, or even within a species, can be misleading if the energy content of the milks vary. A lower concentration of a milk constituent may not indicate that an infant receives less, as the infant may regulate his intake and/or the mother may regulate milk production to enable a similar total amount of the constituent to be ingested despite differing concentrations.

The reader should notice a parallel with our discussion about milk protein content in chapter 6, where we showed that despite large differences in milk protein concentration, the milks of Asian elephants and white rhinoceroses are effectively equivalent as protein sources, once milk energy is taken into account (see figure 6.5). If we were to measure any particular hormone in Asian elephant and white rhinoceros milk we would predict that the absolute concentration was most likely to be highest in the elephant milk. But would that imply that elephant calves ingested more of that hormone each day or even during each nursing bout than did rhinoceros calves? Not necessarily. We would have to take into account the volume of milk ingested, and that is likely to be higher in rhinoceros calves.

Researchers and others primarily interested in the variation found in human milk may wonder if that concern is significant for their purposes. Human milk certainly does not vary among individuals as much as elephant and rhinoceros milk does. However, from our own data we have seen variation in water content of human milk samples from 80 to 89% and variation in milk energy from 0.5 to 1.3 kcal/g. Much of that variation can be explained by stage of lactation, with the low-water, high-energy milk samples representing late lactation. But that still implies that a direct comparison of milk hormone concentrations between samples without accounting for water or energy content may obscure rather than reveal biologically significant differences.

We propose two potential units for expressing hormones in milk that will enable better comparisons between samples, especially for comparisons across species: expressing levels on a per energy basis (units of ng/kcal), and expressing levels per mg of milk protein (units ng/mg). The pattern of EGF across lactation in gorilla milk differed depending upon the units of measure, suggesting that these different units may be telling us different stories. The challenge is to determine which pattern (units) best describes the phenomenon of interest and thus reflects most accurately the biological significance of any variation or lack thereof.

It is disappointing that the literature on human milk bioactive factors usually does not include measures of any of the macronutrients. One study on human milk adiponectin did measure total protein and showed that it was correlated with adiponectin (Ley et al., 2012). We suggest that the interpretation of variation in the concentration of any milk constituent benefits from knowledge of the concentration of macronutrients in milk.

Transmitting Food Preferences

We end this chapter with a brief review of an example of maternal transmission of behavior through milk: food preferences and tolerances. In other words, to what extent can maternal information regarding adult diet be transmitted to her offspring via milk? In addition to feeding offspring does milk guide what the offspring later will eat after they have been weaned?

Food preferences are often learned very early in life. When a parent provisions its offspring it provides information in addition to nutrition. What a young animal is fed by its parents often (but not always) contains cues that may guide what foods the offspring will eat later in life. What a young animal is fed by its parents often (but not always) is what it will eat after it has developed to where it can independently feed. This is true for many vertebrates, but in mammals there is a twist.

Mammals are unique in that the first food for a neonate is always mother's milk. Milk does convey cues regarding maternal diet. Many flavor molecules cross into milk from the mother's blood stream. This has been demonstrated in a variety of species. One of us (MLP) still remembers drinking very funny tasting milk from the local dairy as a child and being told that the cows had gotten into a pasture with a fair amount of onion grass. In research, both onion and garlic have been used to flavor mother's milk. Mammalian mothers fed on foods containing onion or garlic flavor produce milks that smell and taste of onion or garlic. This is true in cows, rats, and humans

(Mennella and Beauchamp, 1991; Beauchamp and Mennella, 2011), and likely true for most mammals. More importantly, flavors in milk have been shown to influence feed preferences and tolerances by offspring later in life.

Review of the Evidence

Baby mammals are capable of showing consistent reactions to tastes and smells. These reactions can be influenced by experiences before and after birth (Cooke and Fildes, 2011). Babies orient more to cloths dipped in their own amniotic fluid, which suggests that sensory information can be passed via amniotic fluid. Garlic odor can be detected in the amniotic fluid obtained by amniocentesis 45 minutes after the mothers ingested a garlic tablet (Mennella et al., 1995). Infants whose mothers habitually consumed garlic during gestation were more likely to orient their heads toward cotton swabs dipped in garlic (Hepper, 1995). In a study of women who planned to breastfeed their babies, the mothers were assigned to one of three groups: the first group was to habitually drink carrot juice during the third trimester and only water during lactation; the second group drank water during gestation and carrot juice during lactation; and the third group drank only water, no carrot juice ever (Mennella et al., 2001). The babies whose mothers drank carrot juice were more tolerant of carrot juice when offered at weaning (about 6 months), with no difference between gestational and lactational exposure. The accepted hypothesis is that flavor molecules from foods eaten by the mother can cross into amniotic fluid and into breast milk.

Research has shown that when rat dams are fed garlic-flavored chow during lactation, the pups prefer garlic-flavored chow after weaning. Again, this was also true for dams fed the flavored diet during gestation. Rat pups preferentially ingested the same diet as their dam ingested during lactation, even if the diet was generally low in palatability based on taste tests with adult rats (Galef and Henderson, 1972). Evidence in rabbits indicates that flavor preferences can be transmitted during gestation (presumably through amniotic fluid) and after, both through milk and fecal maternal pellets (Bilko et al., 1994).

Sometimes what is transmitted might be more properly termed as a tolerance to a flavor rather than a preference. Orphaned lambs were fed milk flavored with either onion or garlic for 50 days and then were offered onion-flavored, garlic-flavored, or unflavored food. The general order of preference was unflavored, onion-flavored, and garlic-flavored last. However, exposure to onion- or garlic-flavored milk increased the tolerance (proportion ingested) for that flavor (Nolte and Provenza, 1992). The lambs did not prefer the onion or garlic flavor; but exposure via milk made them more accepting of it.

Breastfed babies at 5–7 months are more accepting of unfamiliar foods than formula-fed babies (Maier et al., 2008), and of caraway flavor (Hausner et al., 2010); the authors' of these studies hypothesize that the greater acceptance of novel flavors by breastfed babies is due to the greater variation in flavor compounds in breast milk compared with formula. Breastfed children continue to be less picky eaters out to 7 years of age (Galloway et al., 2003), suggesting a long-lasting effect of early taste exposure via milk.

The common marmoset monkey shares an unfortunate trait with humans; marmoset infants are vulnerable to developing obesity (Power et al., 2012; 2013). An obese mom is a risk factor for marmoset pediatric obesity. Another consistent trait found in obese juvenile marmosets is a tolerance for high-fat food. In general, marmosets preferred the low-fat diet to the high-fat diet when offered a choice (Ross et al., 2013). Normal, lean juveniles ate little or none of the high-fat diet. Obese juveniles were willing to consume some high-fat diet, ingesting up to 30% of energy intake from the high-fat food. Milk composition is variable in marmosets, especially fat content (Power et al., 2008). Larger, fatter females produce higher-fat milks than do lean females. Although it is not proven, these findings suggest that exposure to a high-fat milk as an infant leads to a tolerance of high-fat food in older marmoset juveniles.

Perhaps the strongest series of studies on this topic has been performed by Julie Mennella and Gary Beauchamp. In their studies, human babies being fed on formula were randomized to whole protein or hydrolyzed protein (broken down into amino acids); these hydrolyzed formulas tend to taste bitter and sour to adults. Before 4 months of age neonates are accepting of most milks, though infants as young as 1.5 months old exhibited some rejection of the hydrolyzed protein formula if they had never been exposed to it before. Infants at 3.5 years of age who were first exposed to the hydrolyzed formula displayed stronger rejections (Mennella et al., 2011). Interestingly, after one month of exposure to the hydrolyzed formula, all of the infants consumed equal amounts of hydrolyzed formula as cow milk formula. An early study by this group had determined that after 4 months of age, babies without exposure to the hydrolyzed formula display aversive facial expressions and a reluctance to continue feeding, while babies exposed to this formula before 4 months of age are tolerant of it, and may even find it preferable (Mennella and Beauchamp, 1996; 1998). At 7.5 months of age infants exposed to the formula since birth consumed more of the hydrolyzed formula than did infants exposed for only one month before age 4 months, indicating not only a sensitive time period for the development of taste preferences, but a dose effect as well (Mennella et al., 2011). The taste preferences extended to other foods as well, as babies fed the hydrolyzed formula were more accepting of bitter tasting cereals (Mennella et al., 2009).

The evidence is fairly strong. There are likely species-specific sensitive periods during which a flavor is accepted and then later tolerated or even preferred, and more exposure leads to a stronger effect. The time periods for many mammals, including humans, include prebirth, likely due to fetal swallowing of amniotic fluid, and postpartum, from flavor cues in mother's milk (Cooke and Fildes, 2011). Mother's milk can affect taste perception and guide later food choice, in effect transmitting maternal behavior (her own food choices and taste preferences) to her offspring.

Questions and Concerns

There are many more studies that we could have summarized in this chapter, listing many more hormones, cytokines, and other signaling molecules found in milk. This is a relatively new area of milk research; as such it is both exciting and, in our opinion, immature. We have a number of concerns we would like to see addressed. The main questions we have are outlined here: (1) Are researchers measuring the hormones in milk in the most biologically relevant way? We would like to see the variation in hormone concentration more often put into the context of variation of other milk constituents. We are concerned that some reported variation in hormone concentration between populations may simply reflect variation in the major milk nutrients. (2) On what tissues and compartments of the neonatal digestive tract might a hormone be acting? Can they act in the mouth and throat, stomach, or do they need to travel to the intestines? (3) We would also suggest that the neonatal digestive process needs to be taken into account more fully when considering the potential action any hormone might have. Proving a hormone can largely escape digestion in the stomach and small intestine is important; but in addition, the stomach enzyme chymosin means that milk is separated rapidly into curd and whey fractions. These two fractions travel at different rates through the digestive tract, with differing implications for potential action by the hormones in the two fractions. Our standard method of obtaining what is called whey in the laboratory is through centrifuging the milk sample; do we know that this technique results in the same division of hormones into the whey fraction achieved by neonatal digestion? Are we truly measuring the whey fraction of milk?

These and other questions are discussed further in the next chapter, where we review the possibility of milk modifying neonatal gene expression.

Chapter 10

Milk Regulates

There have been many pivotal discoveries and breakthroughs in biology over the last few centuries. One of the most fundamentally important has been the blending of the concepts of evolution with the discovery of the underlying biochemical means by which inheritance of traits is passed through generations. We are truly beginning to deal with the stuff of life. But only beginning. Biological knowledge is progressing rapidly, but it is humbling to realize that we still have a long way to go.

Deoxyribonucleic acid (DNA) was first isolated from white blood cells by Friedrich Miescher in 1869, but it was not recognized as the principle molecule of inheritance for many more decades. In the late 1870s and 80s investigations into structures within cells led to the identification of chromosomes by such researchers as Edouard Van Beneden (1846–1910), Otto Bütschli (1848–1920), and Walther Flemming (1843–1905).

Around 1900 a number of botanical researchers independently rediscovered Gregor Mendel's work: Hugo de Vries (1848–1935), Carl Correns (1864–1933), William Spillman (1863–1931), and Erich von Tschermak (1871–1962). In an amusing coincidence, von Tschermak's maternal grandfather, Eduardo Fenzl (1808–1879) was a botany professor at the University of Vienna who apparently taught botany to Gregor Mendel. Despite this, von Tschermak was unaware of Mendel's publications.

Hugo de Vries coined the term "pangenes" to describe the hypothetical units of inheritance suggested to exist by Mendel's laws. His experiments on wild variants of the flowering plant *Oenothera lamarckiana* led to his conception of a theory of mutation. In 1906 William Bateson used the term "genetics" in a public lecture to describe the study of inheritance. Bateson was a saltationist, in opposition to the gradualism favored by Darwin. The new genetic studies and the mutation theory of de Vries appeared to cast doubt on Darwinian gradualism; but the work of R.A. Fisher, J.B.S. Haldane, and Sewall Wright, among others, largely reconciled the dueling branches into the modern evolutionary synthesis.

In the early 1900s, Theodor Boveri (1862–1915) and Walter Sutton (1877–1916), studying sea urchins and grasshoppers, respectively, independently showed that Mendel's laws applied to the inheritance of chromosomes. This led to the chromosome theory of inheritance (sometimes called the Boveri-Sutton chromosome theory or the Sutton-Boveri theory; there is mild dispute over who should get precedence). In 1913 Eleanor Carothers (1882–1957) provided strong support for this theory by demonstrating independent assortment of chromosomes in grasshoppers. The matter was settled by Thomas Hunt Morgan's (1866–1945) work on inheritance and genetic linkage in *Drosophila* for which he won the Nobel Prize in 1933.

The Russian biologist and geneticist Nikolai Koltsov (1872–1940) attempted to link genetics, embryology, and chemistry (Morange, 2011). In a prescient lecture in 1927, Koltsov predicted that the molecule of inheritance would consist of two mirror-image strands of a giant molecule that would replicate themselves by using each separate strand as a template. Koltsov believed that the chromosomes were composed of that replicating molecule of inheritance, but he incorrectly believed the main structural elements of chromosomes were amino acids in a long polypeptide chain, and that the various nucleotides were not important (Morange, 2011). The results of the Hershey-Chase experiment published in 1952 by Alfred Hershey (1908–1997) and Martha Chase (1927–2003) confirmed that DNA, not protein, was the molecule of inheritance. And finally, in 1953, James Watson (1928–) and Francis Crick (1916–2004), building on work by Rosalind Franklin (1920–1958), published the correct double-helix structure for DNA (Watson and Crick, 1953).

Understanding the structure of DNA and chromosomes is a necessary part of understanding how organisms work, but the important knowledge is how the expression of DNA is regulated. Every cell in your body has the same DNA, but the expression of that DNA differs dramatically between the different cell types. Understanding the various means by which DNA expression is regulated has been what perhaps could be seen as the holy grail of genetics. From the beginning, geneticists understood that learning what determines why a cell expresses its DNA in its particular pattern is the crucial knowledge to understanding how organisms are made and how they function.

In 1949 the existence of a different chromosomal structure found only in nuclei from cells of female mammals was discovered (Barr and Bertram, 1949). In 1959 these Barr bodies were shown to be condensed and heterochromatic X chromosomes (Ohno et al., 1959). The phenomenon of X chromosome inactivation in mammalian females had been discovered. That was the beginning of understanding that the actual physical structure of DNA can change, and those changes are part of how DNA expression is regulated. Since that first discovery scientists have gone on to demonstrate a variety of mechanisms of epigenetic regulation of DNA expression. The important point for this book is that milk contains biochemicals that are capable of effecting gene expression of neonatal cells directly and through influence on the epigenome.

Epigenetic Programming of Development

Mother's milk appears to be able to exert epigenetic effects on neonates. For example, important aspects of uterine wall development occur in the first few days after birth in female piglets (gilts), and these morphogenetic changes are estrogen-receptor-dependent and estrogen-sensitive (Chen et al., 2011). Gilts fed milk replacer from birth differed from gilts that nursed by having no detectable expression of estrogen receptor-α or vascular endothelial growth factor (VEGF) in uterine tissue (Chen et al., 2011). Exogenous relaxin enhanced estrogen receptor-α gene expression in nursed gilts but had no effect on formula-fed gilts, though it enhanced VEGF expression in both. Bioactive factors in colostrum (earliest milk) appear to be necessary for normal gene expression in gilt uterine tissue (Bartol et al., 2008; Chen et al., 2011). Thus, the epigenetic programming of female piglet uterine tissue development is regulated by maternal signals via milk.

Milk may influence the neonate's gene expression and epigenome through milk's effects on the gut microbiome. There are many ways that the microbial ecology of the gut can influence gene expression of the host. Microbes produce a myriad of bioactive compounds that are released into the environment. The microbial composition of the gut microbiome can influence the intestinal pH, its osmotic state, water content, and many other biochemical properties. The microbes also produce molecules that may signal directly or indirectly to host cells. They also produce compounds that have the potential to influence epigenetic programming. For example, some (but not all) *Bifidobacteria*, one of the predominant genera of microbes found in the feces of breastfed infants, produce substantial amounts of folate, which is released into its environment (Pompei et al., 2007). Folate acts as a methyl donor and can facilitate methylation of DNA. Butyrate, a short-chain volatile fatty acid produced by microbial metabolism, inhibits histone deacetylase, and thus enhances histone modifications. The microbiome of breastfed babies produces more acetate and less butyrate than the microbiome of formula-fed babies, implying different epigenetic potential for breastfed versus formula-fed gut microbiomes.

Regulation by Non-Coding RNA

Most of the DNA in our cells is not translated into a peptide. Genes that produce peptides account for only about 2% of the human genome. Instead of the 100,000 genes producing a translated product as predicted before the Human Genome Project, the newest estimate is only about 20,000. But far more of our genome is transcribed into RNA than just the peptide-producing genes. Many transcribed DNA sequences produce RNA sequences that perform functions other than to produce a peptide—for example, transfer RNA and ribosomal RNA genes. Some estimates are that DNA that is transcribed into functional RNA that does not become translated into any product accounts for four times as much of our genome than does the peptide-producing genes.

In the early 1990s, small, non-coding RNAs (lin-4 and let-7) were discovered to play important roles in the timing of development in the roundworm *Caenorhabditis elegans* (Lee et al., 1993; Reinhart et al., 2000). Since that time, our understanding of the importance of small non-coding RNAs in the regulation of gene expression has exploded. Small interfering RNA (siRNA), piwi-interacting RNA (piRNA), and microRNA (miRNA) are all species of RNA transcribed from the DNA that serve to regulate the translation of messenger RNA (mRNA) into a peptide, through both pre- and post-transcriptional mechanisms. Milk contains all these forms of RNA: mRNA, siRNA, piRNA, miRNA, and probably more that have yet to be defined. We restrict our discussion to miRNA found in milk; but other coding and non-coding RNAs may have important functions in milk.

miRNA

MicroRNAs (miRNA) are small, non-coding strands of RNA transcribed from endogenous DNA that regulate gene expression post-transcription by binding to complementary strands of messenger RNA (mRNA) and generally act to repress translation of the mRNA either by blocking translation or facilitating the degradation of the mRNA. They are short (usually 22 base pairs) RNA fragments that are produced from a transcribed segment of DNA that results in a pre-miRNA that has a distinctive hairpin structure.

At least half of mammalian mRNA has been under selective pressure to conserve pairing with miRNAs, implying that the role of miRNAs in regulating gene expression is an important aspect of the regulation of gene expression. Any particular miRNA may be able to bind to mRNA produced by multiple genes, and multiple miRNAs may act on any particular mRNA. Thus, miRNAs represent an extremely complex and powerful evolved mechanism to regulate gene expression.

These small regulators of post-transcriptional gene expression can act as an intracellular effector but also as an intercellular effector. They act within cells to regulate cellular gene expression, but miRNAs are also packaged into exosomes (microvesicles), which can be extruded through the cell membrane and into the extracellular fluid. These exosomes are composed of fragments of the cellular membrane, and thus are protected by a lipid membrane. RNA packaged into exosomes is resistant to RNase and other enzymatic degradation. Thus, miRNA from one set of cells (in a particular organ, for example) enter the bloodstream in a protected vesicle and travel through the circulatory system to be eventually absorbed into another cell. They deliver their contents within the cell, and thus act as an intercellular mechanism for regulating gene expression. Exosomes are able to cross biological barriers, including the blood-brain barrier (Alvarez-Erviti et al., 2011).

Milk miRNAs

Mammalian milk has been shown to contain significant concentrations of a wide variety of miRNAs. Studies have detected miRNA in human breast milk (Kosaka et al., 2010; Zhou et al., 2012), cow milk (Chen et al., 2010), pig milk (Gu et al., 2012), yak milk (Bai et al., 2013), and rat milk (Izumi et al., 2014). Many of the miRNAs found in milk have immune-related function, regulating T-cells and inducing B-cell differentiation, and thus have the potential to activate and potentiate immune function. Immune cell–related miRNAs were by far the most common in cow milk (Chen et al., 2010). About one-third of the known human pre-miRNAs were detected in breast milk, with two-thirds of the well-characterized immune-related pre-miRNAs detected (Zhou et al., 2012). There are non-immune function–related breast milk miRNAs as well, that may regulate other physiological and metabolic processes and development, implying that miRNAs may represent a regulatory mechanism allowing the maternal genome to influence development of her offspring.

These milk miRNAs appear resistant to degradation by acid and RNase digestion (Izumi et al., 2012). Subjecting milk samples to heat, acid, and RNA-degrading enzymes does not affect the measured concentrations of milk miRNAs, strongly suggesting that the exosomes would serve to protect the miRNAs from the acid stomach (Kosaka et al., 2010). This resistance to degradation is likely due to the miRNAs' being contained within exosomes, tiny endosome-derived membrane vesicles that are released by cells into the extracellular environment (Zhou et al., 2012). Microvesicles (likely exosomes) containing miRNA have also been found in cow milk (Hata et al., 2010). Breast milk miRNAs likely survive intact into the small intestine, where they may be taken up by lumen epithelium. However, the exosomes will then be subject to enzymatic degradation from intestinal lipases and proteases. Incubating mouse milk in intestinal fluid does produce a significant decline in measured miRNAs over time (Title et al., 2015).

Human breast milk contained the highest concentration of total RNA of 12 human body fluids tested, almost 3 times higher than seminal fluid, which had the second highest concentration, and more than 80 times the concentration found in amniotic fluid (Weber et al., 2010). The number of miRNAs with known function detected in breast milk ranged from 429 (Weber et al., 2010) to more than 600 (Zhou et al., 2012). Among the 20 most common miRNAs detected in breast milk, 4 were only found in breast milk and not in the other body fluids tested, and 6 were also found in amniotic fluid (table 10.1).

These milk miRNAs appear to be absorbed into circulation. The concentration of circulating miRNAs in piglets was directly and significantly associated with the concentrations in the sow's milk (Gu et al., 2012). A study in adult humans showed that ingestion of cow milk resulted in a 2.5-fold increase in circulating miRNA-29b, one of the most common miRNAs in cow milk (Baier et al., 2014). These findings suggest that milk miRNAs are able to be absorbed into circulation when ingested. The human study even found evidence for change in gene expression in peripheral mononuclear blood cells after milk ingestion, consistent with the expected action of the absorbed miRNA (see end of chapter 12 for more detail). However, a mouse study found no evidence of uptake of two specific miRNAs (Title et al., 2015).

The potential functions of many of the common miRNAs found in milk have been determined by sequence analysis, and for some by direct experiment/observation. Many of the highest concentration milk miRNAs are

Table 10.1. The 20 most abundant miRNAs detected in either breast milk or amniotic fluid in order of abundance

Breast milk	Amniotic fluid	Breast milk	Amniotic fluid
miRNA-335*†	miRNA-518e	**miRNA-515-3p**	miRNA-590-3p
miRNA-26a-2*	**miRNA-335***	[miRNA-513c]	miRNA-873
miRNA-181d	[miRNA-302c]‡	**miRNA-671-5p**	miRNA-410
miRNA-509-5p	**miRNA-515-3p**	[miRNA-490-5p]	**miRNA-509-5p**
miRNA-524-5p	[miRNA-452]	miRNA-367	[miRNA-548d-5p]
miRNA-137	miRNA-892a	miRNA-181b	miRNA-223*
miRNA-26a-1*	**miRNA-671-5p**	miRNA-598	miRNA-616*
[miRNA-595]	**miRNA-515-5p**	**miRNA-515-5p**	[miRNA-148b*]
miRNA-580	**miRNA-137**	[miRNA-578]	miRNA-590-5p
miRNA-130a	[miRNA-593*]	miRNA-487b	miRNA-302d

Data from Weber et al., 2010.

*Indicates the strand that is typically less common.

†miRNAs in **boldface** are among the 20 most abundant for both breast milk and amniotic fluid.

‡miRNAs in [brackets] were only found in either breast milk or amniotic fluid, and not in plasma, tears, urine, seminal fluid, saliva, bronchial lavage, cerebrospinal fluid, pleural fluid, or peritoneal fluid.

common across multiple species. For example, miRNA-21 is one of the most common miRNAs found in human, cow, and rat milk (Munch et al., 2013; Chen et al., 2010; Izumi et al., 2014). miRNA-21 is also linked to regulation of growth and adipogenesis through both upstream and downstream regulation of mTORC-1 signaling (Melnik et al., 2013). Thus, variation in concentration of milk miRNA-21 between dams is a plausible source of variation in both growth and fat deposition in infants.

Milk Fractions

There are three fractions of milk as commonly measured in the lab: cells, the lipid fraction, and skim milk. Milk miRNAs are not distributed uniformly among these milk fractions.

Cells of course contain miRNAs; the lipid fraction also contains high concentrations of miRNAs, not surprising considering that exosomes have a lipid membrane and thus likely associate with the lipid fraction. Skim milk also contains exosomes and miRNAs, but at the lowest concentration. Human milk fat globules have been shown to be enriched in milk miRNAs and to include miRNAs that are at low concentration or absent from the skim portion of milk (Munch et al., 2013), suggesting that variation in milk fat concentration might also result in variation in milk miRNA concentration and proportions of miRNAs. In this study, women who consumed a high-fat diet had consistent changes in the concentrations of certain milk miRNAs; in particular, miRNA-27 and miRNA-67 were enriched several-fold (Munch et al., 2013).

Possible Sites of Action in the Neonate

In order for maternal miRNA in milk to affect neonatal gene expression the exosomes containing the miRNA need to enter neonatal cells. Milk enters the neonate through the oral cavity and passes through the digestive tract. The different regions of the neonatal GI tract offer a variety of possibilities for actions by milk miRNAs. There are no real data to evaluate these possibilities, so the following is informed speculation.

The first possible site of action is the oral cavity. There will be low residence time in the mouth, as the milk containing the miRNAs is rapidly swallowed, but the exosomes will contact cells in the mouth and throat. The oral cavity has strong connections with circulation, with blood vessels in gums near the surface. Microbes in the mouth certainly can gain systemic access, and periodontal disease is a risk factor for many infections throughout the body. Although there may be only short time for exosomes to be absorbed, it is the first opportunity for systemic/circulating action by milk miRNA, and the exosomes will be subjected to minimal digestive processes.

Exosomes that pass through the mouth and throat will arrive at the stomach. Given the moderate pH in the neonatal stomach (pH of 4–5), lab studies suggest milk exosomes and miRNAs would survive. Might they be absorbed by cells in the stomach? The digestive action of the stomach on milk will affect how long the exosomes and miRNAs remain in the stomach before moving through the rest of the gut. We consider some possible complications raised by the neonatal milk digestion in the next section of the chapter.

The indirect evidence strongly suggests that the exosomes will reach the small intestine intact. Here, contact with intestinal epithelial cells provides ample opportunity for absorption into cells, and possibly into circulation. The data from piglets and adult humans shows that milk miRNA do get into circulation (Gu et al., 2012; Baier et al., 2014), and the small intestine is the most likely site for absorption in both these species. However, another study using newborn knockout mice null for miRNA-375 and miRNA-200c/141 who were nursed by wild-type dams found no evidence of uptake of either of these milk miRNAs and evidence of significant degradation in the intestine (Title et al., 2015). Neither of these miRNAs are highly expressed in mouse milk, and they may not be the best choices for testing uptake of miRNAs (Alsaweed et al., 2015). However, this study provides a cautionary note that we do not know for certain whether milk miRNAs are absorbed by infants in biologically significant amounts.

If the exosomes survive into the large intestine they likely have an equivalent ability to affect colonic epithelial cells and possibly absorption into circulation as in the small intestine. They also may have the possibility to interact with the microbiome. We are not aware of evidence to preclude mammalian miRNA having possible effects on microbial gene expression. A speculative question to be considered: Can milk miRNA affect the gene expression from the microbiome?

Curds and Whey Revisited

To understand how milk miRNAs might be able to affect neonatal gene expression, the process of milk digestion must be taken into consideration. The action of the enzyme chymosin in the stomach creates curds, formed by clumping together of casein micelles. The whey will rapidly leave the stomach while the curds will remain behind, to experience fuller digestion in the stomach. Which fraction will milk miRNAs associate with? Probably both; but the curds will contain most of the fat in milk, and in undigested milk the exosomes are more associated with the lipid fraction of milk. Would exosomes be able to escape the curds and leave the stomach, or will they be trapped, to be slowly released through the digestive process? The answer may have consequences for how quickly milk miRNA can affect the neonate's gene expression, and the extent to which they have an effect.

Scientists are measuring many things in milk; miRNAs are some of the newest, and we think potentially most exciting, for understanding possible regulatory functions of milk. But how we measure a substance in milk in the laboratory is not how the infant will experience it. An important consideration that needs to be added to research on the effects of milk on neonates is how the neonatal milk-digesting system affects the concentration, availability, and potential activity of milk-signaling molecules.

Chapter 11

Developmental Origins of Health and Disease

The developmental origins of health and disease (DOHaD) is a major new paradigm for understanding many of the metabolic diseases of modern humans (e.g., diabetes and cardiovascular disease). Vulnerability to disease in adult life is linked to perturbations in development during critical time periods in fetal and neonatal life. These perturbations are caused by environmental signals, often generated by or at least filtered through the mother. What the mother experiences influences what her fetus and later her neonate experiences. The regulation of placental mammalian development depends to a large extent on maternal biochemical signals to her offspring. Maternal biochemical signaling in placental mammals starts during gestation through the maternal-placental-fetal connection; but after birth this signaling continues via milk. Evolutionarily, regulatory signaling via milk is older than the signaling through the maternal-placental-fetal connection. Milk far predates the placenta. We suggest that the importance of maternal biochemical signaling in guiding offspring development is an ancient adaptation in the lineage leading to placental mammals, dating back to the origin of lactation and becoming enhanced with the evolution of the placenta. In this chapter we explore the continuation of signaling from gestation through lactation, examining the similarities and even identical signaling systems. We review the tenets of the developmental origins of health and disease, and consider the extent to which developmental signaling from the mother via milk may program development and metabolism of her offspring.

Early-Life Programming

The range of developmental outcomes arising from environmental effects has different implications for the evolution of these changes by selection. A developing organism that is energy- or nutrient-restricted to an extent that still allows survival but results in a stunted individual may simply represent the best outcome possible given the environmental constraint. The environment constrains more than it guides development in this instance. However, selection undoubtedly still has acted on the developmental program such that under constraint, certain organ systems are spared at the expense of others. Some deficits of function will have greater adaptive consequences than will others, and selection would act to favor deficits with lesser fitness consequences over ones that reduce fitness to a greater extent. For example, in intrauterine growth restriction (IUGR) fetuses redistribute blood flow such that the development of brain, heart, and adrenals are less affected than other organs. The greater head-to-abdominal circumference observed in IUGR infants largely is explained by a greater impairment of liver growth relative to brain growth, leading to a lower abdominal circumference (Nathanielsz, 2006). The organ-sparing effect is relative; in cases of severe IUGR, brain and heart will also be affected, but to a lesser extent than other organs.

The theory behind the development of adult disease due to early life responses to challenges suggests a subtler and more adaptive process. If circumstances during early life are predictive of challenges to be faced in later life then developmental changes in response to early life circumstances could be adaptive for later life, and not merely the best of a bad situation. In this scenario maternal circumstances (e.g., plane of nutrition, disease history, social status) result in biochemical signals to the developing offspring, both *in utero* through the maternal-placental-fetal connection and after birth via milk, which are reliable indicators of later life circumstances. The offspring's development is affected to achieve an eventual physiology that is appropriate to its expected later life condition.

Developmental programming of this type is considered adaptive in the short to medium term (relative to the reproductive life span). However, it can create an increased risk for later disease either because the resulting physiology is not as robust as other outcomes (e.g., allows survival and reproduction but results in frail or unhealthy physiology in late adult life), or because the later life circumstances do not match that predicted from early life experience. The consequence is a mismatch between physiology and environment, which enhances vulnerability to poor health and increased metabolic disease.

Maternal Biochemical Signaling to Offspring

Mothers of all taxa signal biochemically to their offspring. In oviparous taxa (e.g., birds and most reptiles) females deposit substantial resources into large eggs, and from those resources the embryo will develop until hatching. Obviously these maternal resources include signaling molecules as well as the basic nutrients and other building blocks of life. For example, maternally derived steroid hormones are deposited into the egg, and the amounts of these hormones can vary, presumably due to maternal circumstances. Experimental manipulations of steroid levels in eggs (by injecting into the egg) have demonstrated that developmental characteristics of the offspring can be affected. For example, bird egg androgen levels are associated with the length of time to hatching, growth rate of embryos, mortality, immune function, and begging behavior after hatching (Schwabl, 1996; Eising et al, 2001; 2003; Navara et al., 2005). One difference between an oviparous species and taxa with live birth, such as most mammals, is that the time period during which the maternal biochemical signals can be directed at the offspring is severely constrained in oviparous species. A female bird cannot later alter what she has put into her egg.

In a placental mammal, maternal biochemical signaling is a continuous process over an extended period of maternal dependence. It starts at implantation through the maternal-placental-fetal axis and continues postpartum via milk. An extensive amount of development of mammalian fetuses and neonates depends upon signals that originate in the mother. These signals change over time, not only in a programmed way that matches the changing developmental circumstances as the offspring age and mature, but also as maternal circumstances change. Mammalian mothers may be able to alter their developmental signals depending on the maternal environment, or in some cases may have no choice, as the signaling depends upon maternal circumstances. Again, this is possible for an egg-laying vertebrate such as a bird as well, but the time frame for mammals is much broader. For most bird species, the maternal circumstances at the time the egg is produced essentially determine the biochemical composition of the egg, and hence the signals transmitted to the developing embryo. For placental mammals, the maternal circumstances at the time of ovulation may have very little to do with the eventual path of development for those processes that can be influenced by later maternal signaling. There are essential time periods during which development of species characteristics can be modified, and outside of those time periods environmental signals may have little effect on development. For any given trait these time periods may be at any point of gestation or even after birth. For marsupials, most of the signaling occurs after birth and through milk. For placental mammals that

produce altricial young, much of the required maternal signaling must occur via milk. For humans, with our fairly precocial babies, the effects of
milk may be less, and more evident, in preterm infants.

Thus, for a mammal, a perturbation of maternal circumstances at any
point during the extensive period of offspring dependence could have later
life consequences for the offspring. In an experimental example, rat pups
injected with leptin shortly after birth have altered expression of hepatic 11β-
hydroxysteroid dehydrogenase type 2 (11β-HSD2). However, the direction of
the change depends upon maternal circumstances during gestation. Pups
from well-nourished dams showed an increase in 11β-HSD2 expression
while pups from dams that were food-restricted by 30% showed a decrease
in expression (Gluckman et al., 2007). The physiological reaction to a manipulation was dependent upon maternal circumstances, presumably due to
the altered transfer of resources and signals from food-restricted dams to
their fetuses causing different epigenetic changes in their pups compared to
the pups of well-nourished dams.

We argue that the ancient adaptation of lactation was the origin of the
mammalian pattern of substantial maternal regulation of offspring development and that mammals derive from a lineage for which maternal regulation of offspring development was enhanced and more flexible than other
vertebrate taxa. In effect, maternal regulation of offspring development over
an extended period of offspring maternal dependence is a significant adaptive feature of mammalian biology that probably predates the existence of
mammals.

Placenta

Although this is a book about lactation, the topic of DOHaD requires consideration of the placenta and maternal-fetal signaling. The chorio-allantoic
placenta is a more recent adaptation of mammals compared to lactation, orig-
inating perhaps as recently as 100–110 million years ago (Murphy et al., 2001).
Although a relatively recent adaptation compared to milk, the evolution of a
chorio-allantoic placenta increased the extent and sophistication of maternal-
fetal biochemical signaling in mammals. The human placenta connects a
mother and her baby physically, metabolically, and immunologically. Made
from fetal cells, it is the first organ any mammal ever makes. The evolution
of a placenta has influenced multiple aspects of mammalian biology, especially for biological regulation, from metabolism to the genome. The placenta allows biochemical signaling to go from offspring to mother, as well

as the reverse. Mothers and babies are "speaking" to each other at a fundamental biological level from the moment of implantation.

The placenta is, fundamentally, a regulatory organ, producing a wide array of hormones, growth factors, cytokines, immune function molecules, and so forth (Petraglia et al., 2005). These molecules act locally as well as at a distance to affect the mother, the fetus, and the placenta itself, in essence exchanging information among these compartments in order to coordinate the necessary physiological and metabolic processes required to produce a viable neonate. The intrauterine environment has been shown to have a strong influence over growth and development, and to have significant effects on offspring's future health and well-being long after birth. The placenta coordinates maternal physiology with fetal development and even plays a role in stimulating maternal changes that prime the female for caregiving behaviors after birth (Numan et al., 2006). The process of birth is the final cue for the onset of lactation.

Placental influence extended to the evolution of genetic mechanisms as well. The existence of this intimate connection increased the salience and selective pressures inherent in the inevitable to and fro of maternal-fetal and maternal-paternal genetic cooperation and conflict regarding fitness imperatives. Male and female mammals are fundamentally different in their reproductive biology with profound evolutionary implications. The placenta is a primary arena where maternal and paternal genes interact, in both cooperation and conflict. The placenta appears to have allowed or even driven changes in aspects of placental mammal genetic regulation so that they differ fundamentally from that of other vertebrates such as birds or amphibians. For example, imprinted genes appear to be a therian mammal adaptation within vertebrates, with many imprinted genes in placental and marsupial mammal species, but none so far found in reptiles or even in the monotremes (Renfree et al., 2013). Imprinted genes are important to placental development and function; and the placenta as a reproductive adaptation was a selective pressure for the evolution of gene imprinting (Reik and Lewis, 2005).

In placental mammals maternal signaling in earliest life is now accomplished through the maternal-placental-fetal connection. Placental mammals have shifted more of development to in utero life. However, significant development occurs postpartum, and is supported by milk. A developing mammal receives maternal biochemical signals of a wide variety over an extended period. These signals not only serve to guide normal development, but also can be varied by the mother in response to environmental conditions that affect her. The ancient adaptation of lactation has resulted in a lineage (mammals) in which maternal regulation of offspring development has been able to evolve to a heightened degree, with the ability to modify development over time based on maternal circumstances.

The Continuity of Maternal Signaling Pre and Postpartum

Not surprisingly, maternal signaling pre and postpartum often shows continuity. For example, EGF and TGF-β-2 are found in amniotic fluid, which is swallowed by the developing fetus. Thus, in utero the developing gut is exposed to these (and other molecules) of maternal origin that likely have regulatory function and are necessary for the appropriate development of the gut. After birth the signaling continues, now via milk. It is instructive to consider that milk was the original source of most potential maternal signals in early mammals. In marsupials it remains the prime source, as the marsupial placenta is both relatively transient and less developed compared to those of placental mammals. Almost all of the necessary resources and regulatory signaling is delivered to marsupial offspring via milk, from what is essentially a fetal stage through weaning. Functions performed by the placenta are accomplished through milk in marsupials.

The delivery of maternal signals to offspring orally is an ancient adaptation in mammals. After the evolution of the chorio-allantoic placenta, with its enhanced signaling capabilities, the fetus is enclosed within the amniotic sac for a much longer period of development. Fetal swallowing of amniotic fluid allowed the ancient oral signaling to continue as well. The adaptation of lactation may have served as a preadaptation to signaling via amniotic fluid.

Adaptive Consequences of Maternal Control of Development

Maternal control of development presents advantages and challenges to both mother and offspring. Ceding control of certain developmental processes to the mother reduces fetal endogenous requirements by taking more from the maternal system. This might even be required in mammalian embryonic life, as the fetal organs and regulatory systems may not be mature enough and the resources initially transferred into the egg are so minimal that endogenous regulation may not be possible. Although it extends the time period during which mothers transfer resources to offspring, it reduces the daily extent of the resource transfer, spreading the cost over time. Most likely a strong maternal influence on fetal and neonatal development is usually a win-win for both mother and offspring; but it does provide grounds for maternal-offspring conflict. For example, greater maternal control allows a withdrawal of resources due to maternal circumstances, in extreme cases aborting; adaptive for the mother, but not for the fetus. Thus, maternal-offspring conflict has become potentially heightened in mammals, especially as the fetus has acquired a greater ability to influence its mother through placental signaling.

The greatly extended time frame over which mammalian mothers can influence offspring development increases the flexibility of phenotype. The genotype-environment interaction is transduced through the mother to a large extent; and changes in maternal circumstances can potentially affect offspring development at many time points from implantation through weaning. Unfortunately, in the modern human environment maternal circumstances may not be as predictive of the infant's post-weaning nutritional plane as it was in our past. In developing countries this can lead to a mismatch between impoverished maternal circumstances but a consistent higher-calorie intake by her offspring. In developed countries, maternal circumstances have exceeded, in terms of fat mass and plane of nutrition, the evolutionary norm. In the past, very few babies were born to mothers with body mass indices above 30 kg/m^2 or who gained more than 40 pounds during gestation. Gestational diabetes was rare. The signals coming to many modern infants from their mothers may be well outside of the normative ranges under which our species evolved. The infant's metabolism and physiology will still adjust in response to those signals, but the adjustments may no longer represent an adaptive response; or, at least, may not result in long-term good health. The mammalian adaptation of extensive and long term maternal influence on development that we hypothesize began with the evolution of lactation may underlie many modern human metabolic diseases. The modern human pathologies associated with DOHaD may represent the failure of an adaptive response due to an inappropriate environment.

PART IV

Our Mother's Milk

In this last part of the book we examine the interactions of milk with humans and our evolution. In chapter 12 we attempt to recreate the evolutionary history of human milk; how it likely has changed from the milk of the common ancestor of great apes and humans over the last 4–6 million years to become the modern human milk of today. What were the possible selective pressures that guided our modern milk composition, and when, approximately, in our evolutionary history did these compositional changes occur?

The differences in milk composition between humans and great apes are not substantial, not surprising considering our close evolutionary connection and that physically, physiologically, and genetically apes and humans are not so different. At first glance, human milk and ape milk are practically the same. But intriguing differences have emerged as ape milks have been better studied. There are known differences between the milks of people, chimpanzees, gorillas, and orangutans. Human milk has a slightly higher fat content, higher concentrations of antimicrobial factors (e.g., oligosaccharides and immunoglobulins), higher concentrations of growth factors and metabolic hormones (at least for those molecules that have been measured), and a dramatically high level of amylase expression. Human milk does have its own quirks that undoubtedly relate to our evolutionary history, and that can give us insights into the various selective pressures that our ancestors, especially the mothers and babies, underwent in the distant and not-so-distant past.

A fundamental difference between our babies and those of the great apes is the extent of brain growth after birth, fueled, at least in the past, by milk. Many authors have proposed that our milk composition must have changed in response to this new nutritional challenge. We address the evidence for any change in milk composition driven by the needs of our large brains. We also examine the question of milk-brain interaction from the reverse perspective: Did the social and technological changes allowed by our increased brain capacity provide adaptive pressure on milk composition? The large brains of our ancestors allowed different behaviors, some of which produced challenges as well as opportunity.

At the end of chapter 12 we look at a behavior unique to people; drinking the milk from another species. This, obviously, is a fairly recent cultural adaptation in our evolutionary history, but it has had profound effects on our species. Milk is certainly good food. Indeed, milk is one of very few secretions from an animal that is actually evolved to be a food. Even for individuals who lose their ability to digest lactose as adults (most humans) many milk products provide high-quality protein and a good balance of calcium and phosphorus. But milk from any species evolved to support growth and development of the neonates of that species. We consider what challenges in the modern world, with plentiful food and reduced need for exertion, might derive from consuming what is essentially a mammalian growth formula.

Breastfeeding and Public Health

The final chapter of the book builds on the nutritional, immunological, and developmental aspects of milk discussed in the earlier chapters to review the health and disease consequences of breastfeeding. We hope that the preceding chapters will have convinced readers that a mother's milk has the potential to significantly affect her baby's growth and development in almost all ways. Theory says mother's milk should be the best food, providing the most appropriate nutrition, immune function molecules, regulatory signals, and even microbes (and food for them as well). No formula can match what evolution has produced.

But that does not mean that breast milk is always best. There are circumstances where breastfeeding may be contraindicated. For example, maternal disease such as HIV, or rare but serious genetic conditions in the baby, such as galactosemia, will result in breast milk conferring risk of disease and pathology to the baby. The evidence is quite strong that formula-fed babies can thrive. We believe the evidence that mother's milk delivers tangible benefits is strong; but mother's milk is not necessary. A child is not doomed if his mother cannot or chooses not to breastfeed. We hope to present a balanced view, based on science. This is a science book. Science is about evidence, data, and replicable results. Breast milk is an amazing and wonderful biochemically complex food that has great benefit for our infants, but it is not a panacea, and will not cure all ills. Recommendations to women regarding breastfeeding need to be firmly grounded in evidence.

We also discuss some of the social aspects of breastfeeding, both in terms of the psychosocial benefits to mother and child, but also in regard to the cultural and psychosocial barriers that may influence the low breastfeeding rates among many women in the United States. However, the topic of this chapter could easily be a book of its own. We cannot be comprehensive in a single chapter. The topics and examples included are ones we feel are especially important, interesting, and informative, and in which we feel confident in our knowledge.

Chapter 12

Milk and Human Evolution

In this chapter we consider the evolution of human lactation and milk composition. We start within the context of humans as anthropoid primates. Most primates use the same general lactation strategy, as outlined in chapter 7: relatively long lactations, nursing their infants frequently, and producing large quantities of dilute milk. There are a few prosimian primates that diverge from this lactation strategy. These are species that park their babies in nests and thus nurse them infrequently and produce a high-fat milk (see chapter 7). Humans fit the general primate pattern of frequent suckling, a high-sugar, dilute milk, and, at least in the past, a long time before weaning. In addition, however, there is variation in milk composition among primate species that exhibits phylogenetic patterns. For example, human milk is more similar to chimpanzee milk than it is to baboon or macaque milk, but more similar to the milks of those monkeys than it is to the milks of New World monkeys or prosimian primates. Both phylogeny and life history influence the composition of every primate milk, including ours.

We know a great deal about the composition of human milk, from nutrients to immune factors, to regulatory molecules of all types. We continue to learn more every day. We are beginning to get a firm handle on where human milk is now. But where did it come from? What aspects of modern human milk represent novel adaptations evolved in our specific lineage, after we diverged from the great apes, and what aspects represent older adaptations from ancient selective pressures? What did the milk of our ancient ancestors look like? If we traveled back in time and were able to get a sample of milk from one of the early genus *Homo* species, or even further back in time to collect australopithecine milk, how would it be similar and how different from the milk of modern *Homo sapiens*? Can we even begin to hypothesize what a proto ape milk was like or the milk of the common ancestor of Old World monkeys and apes? As usual we provide more questions than definitive answers; but we attempt to address these questions based on the scientific evidence.

Table 12.1. Approximate major nutrient composition of ape and human milk, and extrapolated composition of Australopithicine and early genus *Homo* milk

	Fat (%)	Protein (%)	Sugar (%)	Oligosaccharides (%)	GE (kcal/g)	NPME* (kcal/g)
Orangutan	2.2	0.8	7.5	?	.054	?
Gorilla	2.5	1.1	7.1	1.0	.057	.047
Chimpanzee	2.1	1.0	7.4	1.0	.054	.045
Australopithicine	2.3	1.0	7.3	1.0	.056	.046
Early genus *Homo*	3.2	1.0	7.3	1.0	.064	.054
Modern humans	3.7	1.0	7.2	1.9	.068	.055

*NPME is the energy from fat and lactose, excluding protein and oligosaccharides.

Unfortunately, we will never obtain a sample of milk from our ancient ancestors; but we can develop reasonable hypotheses about what it may have been like by comparing modern human milk with that of our closest relatives, the great apes, and then comparing the similarities between us and apes with how monkey milk is different. The comparative perspective allows sensible hypotheses about the evolutionary path of our milk composition.

All mammalian species have evolved a unique milk. At the same time, phylogenetic constraints exert a major influence on milk composition. Milks of closely related species are similar. Milk composition represents sets of evolved compromises between the current life history strategy of the species and its ancestral conditions. Closely related species generally have similar life history strategies. It is very difficult to disentangle whether milk composition is constrained by genetics which then constrains life history strategy options, or the reverse. Probably both pressures are operating.

In the first section of this chapter we use the principles of phylogenetic constraint in conjunction with the differences in reproduction and infant growth and development between humans and great apes to produce an estimate of the nutrient composition of both australopithecine and early *Homo* milks. Our knowledge of the composition of great ape milk is limited compared to what we know about human milk, but recent research has demonstrated that great ape milk is similar, but not identical, to human milk (table 12.1).

Our Ancestral Milk

We are anthropoid primates. More specifically, we are members of the hominoid lineage of anthropoid primates, along with the orangutans, gorillas, and chimpanzees. About 4–7 million years ago our specific branch of the hominoid bush diverged from the branch leading to modern chimpanzees. The gorilla lineage had diverged from our linage a few million years earlier, and the Asian hominoid lineage, represented by the orangutans, had split off around 12 million years ago. The nutrient composition of great ape milks and modern human milk is given in table 12.1, along with hypothetical compositions for australopithecine and premodern *Homo* milks. The milks of the great apes are probably the best model for milk from an australopithecine mother—low in protein, low in fat, high in sugar and water. Such a milk would have fit a lactation strategy of frequent suckling of the infant over a long lactation period, probably lasting 3–5 years. Growth and development would have been slow compared to that seen in monkey species, just as in the modern great apes.

While australopithecine milk most likely resembled great ape milk, would milk composition have changed during or after the transition from an australopithecine to early genus *Homo*? Certainly modern human milk differs from great ape milk in having a somewhat higher fat content, although sugar and protein content are essentially identical. There are other differences between human and ape milks, which we discuss in detail later in this chapter; but most of those differences are among the non-nutritive aspects of milk, and likely evolved much more recently. What certainly happened is that the fat content of milk in genus *Homo* increased over time.

There are a number of reasons why higher milk fat might have a selective advantage. A low-fat, high-sugar, high-water milk would require frequent nursing bouts, since the volumes of both the mother's mammary glands and the babies stomachs would restrict the total amount of milk that could be transferred during any given sucking event. The prosimian primates with higher-fat milks park their babies in nests and then leave them for considerable periods of time while the mothers forage and engage in other selectively advantageous activities (Tilden and Oftedal, 1997). They cannot nurse their young as often as a monkey, ape, or human who carries their baby around with them. The solution, similar to that evolved in other species such as rabbits and tree shrews, which suckle their offspring infrequently (see chapter 7), is to produce a nutrient-dense milk—a high-fat, low-sugar milk. We can confidently exclude this hypothesis as an explanation for the higher fat in human milk. Human mothers suckle their babies often; no differently, in general, than the mothers of other anthropoid primates, including ape mothers. There is no evidence for a reduction in the frequency of suckling behavior in our ancestors that would have selected for a higher-energy milk.

Note that it isn't completely ludicrous to consider this possibility. An evolutionary "just-so" story could be devised in which the mothers of our early ancestors left their babies in a communal camp (crèche), perhaps looked after by older females (grandmother hypothesis) while the mothers went out and foraged for food to be brought back to the camp. Nursing frequency would have decreased, and milk composition might have changed in response. There just isn't any evidence this happened, and there's plenty of evidence that modern hunter-gatherers do not use this potential strategy. Further, only milk fat appears to have increased. If milk composition had changed in response to a decrease in suckling frequency then protein content would be expected to have increased in concert with increased fat and decreased sugar content to allow a decrease in water content.

The obvious biological change between the australopithecines and early *Homo* is the dramatic increase in brain size in our ancestors. In 1995 Aiello and Wheeler proposed that the human brain, due to its metabolic expense to grow and maintain, presented an adaptive challenge to early members of genus *Homo* (Aiello and Wheeler, 1995). The expensive tissue hypothesis in effect proposed that the Australopithecine dietary ecology was not capable of supporting this extra brain matter. To afford a larger brain our early ancestors had to shift to a diet of higher-quality foods, which means foods that provide more energy (and possibly other particular nutrients) per gram. One aspect of a food that generally lowers its quality is being difficult to digest. Leaves generally provide a lower calorie-per-gram return than do fruits and meat, in large part because they are more difficult to digest due to the plant cell wall fraction (called fiber). In practice, the dietary shift that *Homo* made early in its existence was to a diet higher in animal matter. This dietary shift, combined eventually with fire for cooking and other methods of food processing that increased food digestibility, enabled our digestive tracts to become less substantial, in effect matching the extra cost of our brains with a decrease in resource needs by our guts. Our brains got bigger, our guts got smaller, and genus *Homo* prospered.

The expensive tissue hypothesis is an explanation for adults. It proposes a scenario by which a large adult brain can be supported. It does not address how to grow that large brain in the first place. If an adult diet change was required to support a large brain, does that imply that the infant diet needed to change as well? Did breast milk of early members of genus *Homo* evolve a higher-fat content due to selection to support enhanced brain growth?

Milk and Brains

Modern human brains grow more rapidly than ape brains during the fetal period; human newborns are born with a brain that is similar in size to that of adult chimpanzees. However, this is only 25% the size of an adult human's brain. Human brain growth continues to be rapid postpartum, especially during the first 18 months of life, a period of time when breast milk is a major source of energy and nutrients and might have been the sole source for the infants of our earliest ancestors in genus *Homo*. The human infant brain accounts for a much larger proportion of energy requirements and of the nutrient requirements of early growth than does the brain of a baby chimpanzee or gorilla. Brain growth in human neonates exceeds the growth of most other organs and tissues. A neonatal human is growing more brain than brawn.

This is not to say that chimpanzee baby brains aren't growing while the infants nurse. All mammal neonates grow their brains fueled to some extent by milk. Some altricial species, such as many carnivores, produce undeveloped neonates whose brains must grow extensively during lactation. Marsupial babies grow almost all their brains on milk. Humans are still the champions, for the simple reason we have such large brains in proportion to our body size. Human brains grow more quickly in utero than do ape brains, with a mean brain weight at birth of 250–350 grams, as noted above, already the size of an adult chimpanzee brain. The rate of brain growth remains roughly the same as the in utero rate for the first year of life postpartum, reaching a weight of 1 kg by around 1 year (Dobbing and Sands, 1973). Thus, human neonatal brain growth averages about 2 g per day for the first year of life. In the past this growth was mostly if not completely fueled by mother's milk.

Brains are expensive. Brain tissue is costly both to produce and to maintain. Of course, this is true of most organs; heart, liver, kidney, and guts are not cheap. They require substantial energy to grow and maintain, as well. But brain is still somewhat special. For one thing, brain is a naturally high-fat organ. The neurons in brain are sheathed with lipid-rich myelin, which acts as electrical insulation and increases the speed of neural transmissions. This is true of all brains, not just human ones. But of course the large relative size of the human brain means that there is proportionately greater requirement for lipids to grow a human brain. Although myelination of neurons begins in the 14th week of gestation, most myelination in human neonates occurs after birth, fueled by milk.

The higher fat content of human milk could be a response to the enhanced brain growth fueled by milk. The extra fat would provide energy, both for maintenance of the brain and for the energetic cost of growing the brain. Milk fat would also be deposited in the brain. The lipid requirement of brains is not for just any fat. There is a requirement for some fatty acids that are important constituents of brain tissue. Palmitic acid and oleic acid are important constituents of brain phospholipids. The long-chain polyunsaturated fatty acid (LCPUFA) docosahexaenoic acid (DHA) is an important constituent of membranes surrounding neural synapses as well as in the retina. Several authors have suggested that DHA may be a limiting or critical nutrient for brain development with implications for human evolution and milk composition (e.g., chapters in Cunnane and Stewart, 2010).

A human neonate will be born with a store of DHA and other essential LCPUFA, transferred from the mother through the placenta; but this store will be sufficient for requirements for several months, not all of infancy. Milk DHA and endogenous production from alpha-linolenic acid (ALA), the precursor to DHA, is required. The mammary gland does not synthesize DHA, and ALA is an essential fatty acid that must be obtained from diet. The sources of DHA and ALA in breast milk are diet and maternal body stores.

Has human milk evolved to meet this hypothesized higher requirement for DHA? Not obviously. Human milk is not different from non-human primate milk in DHA (Milligan and Bazinet, 2008; Milligan et al., 2008b). Milk DHA content in all species, including humans, is quite variable, with the variation generally assumed to reflect maternal diet. Fish and other seafood are excellent sources of DHA, a key component in the theory concerning the importance of an aquatic dietary niche for early human evolution (Cunnane and Stewart, 2010; Cunnane and Crawford, 2014). Dietary intake of DHA will increase milk DHA content, and maternal intake of DHA during pregnancy and lactation is associated with several positive outcomes in human neonates, as well as is DHA supplementation of infants. The human mammary gland has not appeared to have evolved any particular adaptation to boost DHA in breast milk, and some authors, including us, question whether dietary DHA is a limitation on human brain development under natural conditions.

How Much Fat Does a Baby's Brain Need?

In the first year of life a human infant's brain gains about 2 g per day, going from about 300 g to 1 kg (Dobbing and Sands, 1973). How much of that mass gain is fat? How much is DHA? These are important questions to consider when evaluating the requirements of human brain growth relative to breast milk. In reality, the amount is pretty low, compared with the amount of milk fat likely consumed per day. Below is a rough, simple calculation to estimate the magnitude of both brain lipid requirements and lipid provided to a baby via breast milk.

Many authors state that brains are high-fat; even we have done so (Power and Schulkin, 2009). But that is a relative comment, and statements that brains are 30–50% fat are providing values on a dry matter basis; that is, without the water. Brains, like all living tissue, are mostly water; especially neonatal brains. At birth a human baby's brain is almost 90% water (Widdowson and Dickerson, 1960; Dobbing and Sands, 1973) and a human adult brain is at least 75% water (Forbes et al., 1953; Widdowson and Dickerson, 1960). Between birth and one year of age a human infant's brain decreases in water content to about 83% and then levels off (Dobbing and Sands, 1973). Based on the assumptions of a brain mass at birth of 300 g (88% water) growing to 1,000 g at one year of age (83% water), with brain dry matter consisting of 50% lipid and DHA comprising as much as 50% of the fatty acids, then the following values can be calculated: Total brain mass increase = 1.9 g/day of which about 1.55 g/day is water and 0.35 g/day is solids; brain fat gain is about 0.18 g/day and DHA gain is 0.09 g/day. Based on breast milk with 4% fat and 0.34% of the fat as DHA (means for human milk across many studies) and an intake of 800 g/day of breast milk by the baby, then the mother delivers 32 g/day of fat and 0.11 g/day DHA to her baby. Most of the fat would be metabolized for energy; but there would still appear to be sufficient milk fat to supply brain requirements, though the DHA amount is close to the estimated requirement. As stated, human milk can vary widely in DHA content due to dietary variation, and values range from 0.06 to 1.4% (Milligan and Bazinet, 2008). Women with below-average DHA content of their milk may not be transferring sufficient DHA to their babies. Also note that if human milk fat content was the same as that of the great apes (2.5% at best) DHA intake from breast milk would likely have been below the estimated requirement. It doesn't look like human milk evolved any special ability to concentrate DHA, but the higher fat content does assist in assuring adequate DHA transfer to the neonate.

Fat Babies

Human babies are born fat, at least relative to most other mammalian neonates (Kuzawa, 1998). Body fat of human neonates is usually between 10 and 15%; most other anthropoid primate infants appear to have much lower amounts of body fat at birth. For example, baboon neonates have under 5% body fat (Kuzawa, 1998). However, baby common marmosets, small New World monkeys, appear to be born with body fat content similar to humans (Power et al., 2012), so the uniqueness of human baby fat among primates is not assured. In any case, there are advantages to being a fat neonate. Baby fat buffers the infant through the initial onset of lactation, when the mammary is producing colostrum with high immunoglobulin content but lower energy content. Baby fat may also provide much of the required fatty acids such as LCPUFAs for immediate postpartum development.

The human placenta must be capable of supplying the necessary energy and building blocks, especially arachidonic acid (AA) and DHA, to build a brain more than twice the size of a chimpanzee neonate's. The human placenta efficiently transports AA and DHA from maternal circulation to the fetal compartment. Concentrations of AA and DHA in cord blood are roughly double the concentrations in maternal circulation (Duttaroy, 2009), implying the existence of an active transport system. Human infants likely are born with significant stores of essential fatty acids, including AA and DHA, which can be applied to brain growth postpartum. It can be argued that much of the LCPUFA requirement for neonatal brain growth and eye development was provided in utero. But that prepartum store of LCPUFAs is unlikely to last for more than a few months. Eventually milk will become the predominant source.

At present there is no evidence that the higher fat content in human milk is apportioned any differently in terms of fatty acid composition. There are some suggestions that milk fatty acid composition differs between monkeys and apes. Monkeys (represented by marmosets, lion tamarins, and howler monkeys from the New World and macaques from the Old World) have significantly higher concentrations of medium-chain fatty acids in their milk, specifically caprylic acid (also known as octanoic acid; 8:0) and capric acid (also known as decanoic acid; 10:0). But there were no taxonomic differences in LCPUFAs among the primates studied; differences in ALA and DHA content of milk were related to diet, not phylogeny (Milligan, 2007; Milligan et al., 2008b).

The increase in milk-fat content in milk of modern humans compared to that of great apes solved two essential developmental challenges: fueling the energy-intensive aspects of brain growth and maintenance, and providing a larger quantity of AA and DHA to the infant even if the concentrations of these LCPUFAs did not change. Although not proven, the hypothesis that milk composition of genus *Homo* evolved to be higher in fat as an adaptive response to support the greater brain growth is viable.

Fat in Place of Sugar?

If the increase in milk fat was related to the larger brain that is the hallmark of genus *Homo*, then the change in milk fat possibly was an early adaptive response and arose perhaps 1 to 2 million years ago. However, we suggest that some of the milk-fat increase may have occurred more recently, due to a change in human milk sugar composition.

The total sugar content of human milk is the same as it is in the great ape milks, at between 7 and 8% (table 12.1). However, the relative proportions of sugars in human milk differ from the ape milks. Human milk contains a much higher concentration of oligosaccharides other than lactose. As much as 20–25% of the sugar in human milk is in the form of what are called human milk oligosaccharides (HMO), more than four times as high a concentration as is found in ape milks (Goto et al., 2010; Tao et al., 2011). These oligosaccharides are generally not digestible by the infant and thus do not provide the infant energy. The hypothesized functions of these HMOs is discussed in detail in chapter 8. Briefly, they perform both prebiotic and antibiotic functions, with some HMOs being metabolized by symbiotic gut microbes, while other HMOs serve as decoy molecules to which potential pathogens attach and are then carried through the gut to be defecated.

The hypothesized adaptive reasons why human milk contains such high concentrations of oligosaccharides is discussed below. One consequence of the high concentration of oligosaccharides is that human milk has less metabolizable energy in the form of sugar than do ape milks. In human milk, at least 20% of the potential energy from sugar is in the form of oligosaccharides as compared to perhaps 5% in ape milks. This means that in 100 g of ape milk there is about 4.5 kcal more metabolizable energy available from sugar compared to human milk. That is the equivalent of the energy available in 0.5g of milk fat. Thus, an increase in milk fat by 0.5% would balance the loss of available energy due to the shift to a higher milk oligosaccharide content.

We don't know when during our evolution milk oligosaccharide content increased. If it was in response to increased exposure to infectious agents, as described in the next section, then the increase was probably late in our evolution, after our ancestors already had big brains. If part of the higher milk fat content of human milk derives from replacing the energy lost by reducing lactose and increasing other oligosaccharides, then the change is a fairly recent one. Still, the difference in milk fat between human and ape milks (1–2%) is more than would be needed to simply replace the energy lost in lactose. In that case early *Homo* milk may have been intermediate between ape and modern human milk in fat content (table 12.1).

Dirty Babies?

Human milk doesn't just have a high concentration of oligosaccharides, it also has a greater diversity. A human milk sample will have hundreds of HMOs, and samples from different women differ in the HMOs expressed (Bode, 2012). Milk oligosaccharides have prebiotic and antibiotic functions; indicative of the importance of the human gut microbiome and of fighting off potential pathogens to human neonatal survival and health.

Interestingly, human milk appears to be exceptionally well endowed with many anti-pathogenic factors, not just HMOs, implying a high evolutionary pressure on infant immune function. For example, human milk has very high concentrations of secretory IgA (SIgA); the level of SIgA in mid-lactation human milk exceeds the concentration measured in colostrum of rhesus macaques (Milligan, 2007). Human milk also appears to have a higher concentration of lactoferrin, a molecule that sequesters iron. Milk is normally low in iron; but in response to infection, lactoferrin concentrations will increase further, lowering iron concentration even more. Since iron is a limiting nutrient for many pathogenic microbes, this will result in starving the invader. Not surprisingly, human milk is low in iron.

The human placenta also appears to have evolved a greater capacity to transfer immunoglobulins. Not all anthropoid placentas appear equal in the ability to transfer IgG to the fetus (Coe et al., 1994). Newborn squirrel monkeys (a New World anthropoid primate) had IgG levels equal to only about 35% of maternal values. In contrast, in the rhesus monkey and chimpanzees IgG levels in newborn infants were roughly equal to maternal levels. The highest levels of IgG were found in human babies, where newborn infant values were actually significantly higher (139%) than maternal values, supporting the finding of an active transport system for IgG in human placenta (Kohler and Farr, 1966). In a prosimian primate, the galago *Otolemur garnetii*, IgG levels in newborn infants were barely detectable at 3.5% of maternal levels (Coe et al., 1994). Of course the galago has an epitheliochorial placenta, and thus would not be expected to have immunoglobulin transfer pre-birth.

There certainly is evidence that maternal transfer of immunoglobulins has been increased in the human lineage. Transfer of both placental and milk-borne immunoglobulins is greater in humans. Both in utero and postpartum human infants are receiving higher transfers of maternal immunoglobulins (IgG and IgA) than do infants of our nonhuman primate relatives. Why should an increased capacity for immunoglobulin transfer in human beings have evolved?

The relatively recent (in an evolutionary sense) advent of agriculture and domestication of animals in human history provides a possible answer. Agriculture allowed an increase in population density and in long-term occupation of living sites. Settlements became permanent, with multiple generations living in the same place. There were more possible vectors for disease (for example, other humans). Without modern sanitation, more people living in the same space meant an increased concentration of human waste, increasing the concentration of parasites and pathogens in the immediate environment. Agriculture also attracted pests, such as rodents and cockroaches, which carry parasites and diseases. And of course as people domesticated animals they also brought those animals' diseases and parasites into the human community. Most new diseases that affect modern humans come from other species: bird and swine flus, HIV, Hantavirus, trichinosis, and toxoplasmosis are all examples of diseases that affect humans but originated in other species.

In many cases, infants will be the most susceptible to and suffer the greatest morbidity and mortality from disease and parasites. The evidence of our placenta and our milk suggests that pathogens affecting infancy were significant selective forces in our evolution. The high transfer of IgG across the placenta, the high fatness of human babies, and the high HMO and SIgA content of milk all may have been adaptive responses to the increase in pathogen load our neonates were exposed to due to the dramatic life-style changes that modern humans were able to achieve thanks to our large, creative brains.

Infant Survival

Mother's milk was vital for the survival of human neonates. Nowadays we have the technology to produce milk replacer formulas that are sufficient for human babies; but the epidemiological evidence still indicates that mother's milk has advantages (see chapter 13). In our past, human breast milk was a necessity for neonatal survival. Older babies are another matter, however. The evidence does not support a strong benefit for exclusive breastfeeding after 6 months of age, and there are some risks (e.g., vitamin D insufficiency and anemia from low-iron status; see chapter 5). By one year of age supplemental foods are considered necessary for human babies.

The weaning process for human babies occurs over many months, even years in some cultures. Human babies survive, grow, and develop on a mixture of mother's milk and other foods for a considerable time period. Although babies may breastfeed for several years, up to three or more in hunter-gatherer communities, all human babies are introduced to supplementary foods at a relatively early age. The recommendations from modern medicine are for exclusive breastfeeding through 4–6 months of age, at which time cereals and other supplementary foods may be given. In many human cultures supplemental foods are introduced very early on, sometimes within the first few weeks after birth.

Six months of age is young for a great ape baby to eat solid food. Wild chimpanzee and gorilla babies generally are not seen to eat solid food before one year of age. Of course in captivity baby great apes do begin eating food earlier. Modern food technology allows human (and captive great ape) babies to begin to eat at an early age. Early weaning foods were probably an important aspect of human evolution. The sooner a mother could begin to shift some of the nutritional burden of feeding her baby from her own body onto other foods that possibly could be provided by other people, the shorter the interval between births. Humans in what are called natural fertility populations, whose lifestyle more closely reflects the premodern technology human lifestyle, on average have interbirth intervals of about 3 years. In contrast, wild gorillas and chimpanzees do not fully wean their babies until 4–5 years of age, and generally have interbirth intervals longer than 5 years. Human population growth is intrinsically higher than that of the great apes, even without modern technology.

Mother's milk may be vital to human neonates, but we suggest that for a considerable length of our evolutionary history, supplemental foods have been vital for human babies from an early age. We hypothesize that provision of supplemental foods to young infants was a successful strategy to increase infant survival, reduce nutritional stress on mothers, and reduce interbirth intervals in our ancestors.

Amylase in Milk

Most of the supplemental foods offered to babies are cereal-based, and thus starchy. The enzyme amylase is necessary for efficient digestion of starch. Amylase is found in both the oral cavity (produced by the salivary glands) and in the small intestine (produced by the pancreas). Salivary amylase and pancreatic amylase are the products of different genes. The salivary amylase gene in primates arose about 40 million years ago from a duplication of the pancreatic amylase gene. The pancreatic and salivary gene promoters are unrelated to each other; the salivary amylase gene promoter appears to derive from insertions by transposable elements, one of which is an endogenous retrovirus (Samuelson et al., 1996). Thus, salivary and pancreatic amylase expression can be independent.

In humans, the salivary amylase gene has undergone multiple gene duplication events. People express multiple copies of the salivary amylase gene, and gene copy number is highly variable (Perry et al., 2007). For example, one person may have inherited 6 copies of the salivary amylase gene from his mother and 7 copies from his father, for a total of 13 copies. Another person may have more gene copies (20 or more) or far fewer (as low as 3). Variation in the concentration of amylase in human saliva is in direct proportion to gene copy number (Perry et al., 2007).

Intriguingly, the salivary amylase genes are also expressed in the mammary gland during lactation. Human breast milk contains salivary α-amylase and has the enzymatic capacity to reduce starch into simple sugars (Lindberg and Skude, 1982). Human breast milk varies in amylase activity between individuals (Hegardt et al., 1984), suggesting that the variable amylase gene copy number has a similar effect in saliva and breast milk.

European and US public health officials recommend combining feeding starchy supplementary foods (e.g., cereals) with a nursing bout, or even mixing breast milk with the cereal to aid digestion. European recommendations allow cereal to be given to babies by three months of age, provided the mother is breastfeeding and feeds the cereal in conjunction with a nursing bout. Milk amylase may have been an important adaptation allowing human infants to be provided with starchy supplemental foods at an early age.

Great ape infants are nursed by their mothers for three to five years; this overlaps time periods when the infant is eating foods that may contain starch. Is the high concentration of amylase in human milk an enhancement of an existing condition, or is milk amylase a unique adaptation of

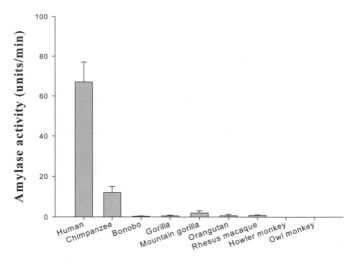

Figure 12.1. Amylase activity in milk from humans and non-human primates.

genus *Homo*? We assayed human milks and milks from nonhuman primates including New World monkeys, Old World monkeys, orangutans, gorillas, chimpanzees, and bonobos for amylase activity (M. Power, unpublished data). Amylase activity was measured using a calorimetric technique that assesses the ability to break down starch. This method cannot distinguish between enzymes that would digest starch, but has the advantage that differences between species in amylase protein amino acid sequence are irrelevant.

As expected, values for human milk amylase activity were high and variable (18–141 units/min). New World monkey milk had no detectable amylase activity; rhesus macaque showed a very low level (less than 1 unit/min) that is difficult to interpret as actual amylase in the milk. Among most of the great apes (bonobo, gorilla, and orangutan) milk amylase activity was similar to the monkey milks, displaying very low levels of potential amylase activity (bonobo less than 1 unit/min; gorilla 0–3 units/min; orangutan 0–3 units/min). Chimpanzee milk samples, however, displayed a significant level of amylase activity (8–20 units/min), although lower than found in human milk (figure 12.1). The chimpanzee data are consistent with expression of two copies of the amylase gene, and the results for human milk amylase activity imply expression by 3 to 20 gene copies.

Most of the data support the hypothesis that milk amylase is a genus *Homo* adaptation; but the data from the chimpanzee milks, of course, rejects the hypothesis. The lack of amylase activity in the milk of bonobos implies either that species has lost mammary amylase expression or chimpanzees independently gained the ability to secrete amylase into milk. These data cannot distinguish between human milk amylase being an enhancement of a trait that existed in the *Pan-Homo* common ancestor, or independent acquisition of mammary amylase gene expression in *Homo* and *Pan troglodytes*.

We speculate that the most likely scenario is that expression of the salivary amylase gene by the mammary gland arose in the common ancestor of *Homo* and *Pan* after the split from the lineage leading to gorillas. We hypothesize that milk amylase initially had relatively little if any adaptive function. Possibly it was an advantage to older infants that began eating plant foods; but our supposition is that it survived by chance and due to a relatively low cost imposed by expressing some amylase in milk. Bonobos lost mammary expression by chance, with little if any selective advantage or disadvantage. It has survived by chance in chimpanzees.

In our ancestors, milk amylase provided a significant selective advantage. Human neonates produce little amylase, either in salivary glands or the pancreas. Salivary amylase production does increase such that by 6 months of age it is close to adult levels, but pancreatic amylase production is still low, hence the generally accepted advice to wait until around that age before introducing cereals. Both salivary amylase and milk amylase can survive the acid conditions of the stomach and become enzymatically active in the small intestine (Lindberg and Skude, 1982), and thus could assist the infant in starch digestion.

The duplications of the salivary amylase gene have been dated to 300,000 years ago. Fire and some form of cooking was an attribute of our ancestors well before that time. Starchy foods, for example wild grains or tubers, could have been significant dietary items. We suggest that starchy weaning/supplemental foods for infants were a successful cultural adaptation that increased infant survivorship, shortened time to weaning, and thus shortened interbirth interval. The ability to give infants supplemental foods also could reduce nutritional stress on mothers, enabling them to regain condition more quickly and become fertile, increasing female reproductive potential. Mothers who fed their older infants starchy foods and breastfed less frequently may simply have regained fertility faster by reducing the contraceptive action of constant nipple stimulation (lactational amenorrhea). Enhanced expression of salivary amylase due to the gene multiplication may have served to support the strategy of feeding starchy supplemental foods to young children, giving youngsters (and adults) additional ability to digest starch. Expression in milk would have enabled even young infants to be fed starchy supplemental foods in conjunction with a nursing bout, easing nutrient demands on the mothers and likely increasing infant survival. We suggest that the adaptive potential of the salivary amylase gene duplications in early humans is best understood in the context of advantages to the nutritional status of infants and young children, more so than in relation to the advantages to adults.

Intestinal Growth Factors in Human versus Ape Milk

If human infants and children were more reliant on foods other than mother's milk compared to wild apes, it is reasonable to consider whether that might imply a faster maturation of the digestion tract. The digestive challenges of milk are different and generally much less than the challenges presented by other foods. After all, milk is made to be digested.

Human milk contains biologically significant concentrations of growth factors thought to be important for digestive tract growth, maturation, and health. Milk from one of our closest relatives, the gorilla, contains physiological concentrations of at least two of these biologically important hormones (EGF and TGF-β2) for an extended time during lactation. The stability of the concentrations of the measured bioactive factors during established lactation, for years in the case of the gorilla, indicate that maternal signaling via milk may be important well beyond neonatal life, and into later infancy.

Human studies indicate that, on average, the concentrations of the measured hormones in human milk are higher than those found in ape and monkey milks (figure 12.2), even after accounting for the slightly higher energy content of human milk. Milk EGF concentrations in gorilla milk (29.7 ± .9 ng/ml [Power et al., under review]) were below those found in mature human milk from western women living in industrialized nations (75 ± 12 ng/ml [Dvorak et al., 2003]), but similar to values found in milk from women in rural Philippines from either a hunter-gatherer tribe or a rice-farming tribe (18.1 ± .8 ng/ml; [Bernstein and Dominy, 2013]). Human milk was consistently higher in TGF-β2 compared to gorilla milk regardless of the human population (Power et al., under review).

Because human milk is consistently higher in fat content (and hence energy) than ape milks so far analyzed, this difference is somewhat amelio-

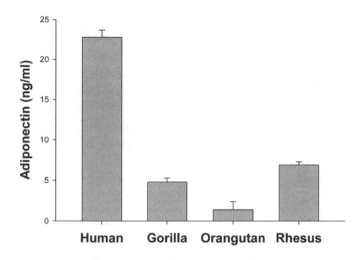

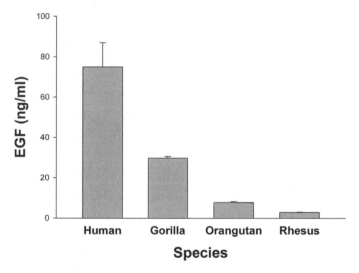

Figure 12.2. Adiponectin and EGF in human and non-human primate milk.

rated when the hormone concentrations are expressed on a per energy basis. However, the values from human milk are still higher, with gorilla milk EGF 47% of human mean values on an energy basis, implying that human infants are exposed to about twice as much EGF from milk even accounting for a possible increased milk volume consumed by ape infants to achieve the appropriate metabolizable energy intake from milk.

This finding does come with an important caveat. Most of the studies of human milk have been done on women in industrialized nations. The range of concentrations found in different human populations indicates that many factors both internal and external to the mother likely affect the concentrations of these hormones in her milk. The fact that women from poorer populations have lower milk concentrations of EGF (but not TGF-β2) compared to western women raises the possibility that the values we are finding in modern western women for some bioactive factors may be outside of the evolutionary norm for humans in the past. At least for EGF, the concentrations in milk are not different between captive gorillas and orangutans and women from rural Philippines whose stature, body condition, nutrition and disease history, and so forth differ markedly from western women. Of course captive great apes may be more analogous to western women, as the captive environment provides more food and less disease and other challenges than these apes experience in the wild. We hypothesize that the concentrations of these hormones in wild gorilla and orangutan milk may be lower than we found in our captive specimens, consistent with the human data. This is an empirical question that could be answered, though not without great difficulty.

The detection of receptors for both EGF and TGF-β2 in gorilla milk (Power et al., under review) raises interesting questions. The ratio of EGF-R to EGF in gorilla mid-lactation milk samples was remarkably stable, suggesting a strong connection between the expression of these two genes by the mammary gland. What role EGF-R may be playing is uncertain. It may be acting as a binding protein, either regulating the action of EGF or possibly providing protection of EGF from stomach acids and digestive enzymes. EGF and EGF-R may be acting as a complex to produce signaling by binding onto infant intestinal epithelial cells, a process known as transsignaling, where a soluble receptor-ligand complex binds to cells to allow signaling even in the absence of a cellular membrane-bound receptor. This is known to occur for IL-6 and its soluble receptor sIL-6R (Rose-John, 2012).

Higher concentrations of growth factors important for intestinal development (e.g., EGF and TGF-β2) may relate to a faster development of the human infant gut, allowing earlier introduction of supplemental foods. The ability to feed infants supplemental foods at an early age (due to the advent of cooking, other means of food processing, domestication of grains, and so forth) could have had profound effects on the survival and fecundity of mothers as well as their children by providing an alternative nutritional resource for infants. But supplemental foods (especially those of long ago) are not as easily digestible as breast milk, and early supplementation might have produced selective pressures to speed up intestinal maturation. That selective pressure may be evident in the increased expression of growth factors in human milk compared to their concentrations in ape milk.

Other Species' Milks

Human beings are the only vertebrates that feed on the milk of other mammals. In relatively recent history (by evolutionary standards), many groups of people began consuming milk and products from milk of other species. This behavior has led to many fascinating changes in us and in other species. For example, it led to the coevolution of ruminants and humans. We have changed cows, sheep, and goats from their ancestral condition, in part in order to get their milk. But some human groups that relied heavily on milk and dairy products have had their genetics change in response to that cultural adaptation, resulting in the ability to produce lactase, the enzyme that digests lactose, as adults.

Milk is certainly good food. Indeed, milk is one of very few secretions from an animal that is actually evolved to be a food. Even for lactose intolerant adults many processed milk products can provide high-quality protein and a good balance of calcium and phosphorus. But milk from any species evolved to support growth and development of the neonates of that species. Milk evolved as a signaling mechanism as well as a nutrition delivery device. Consumption of milk and milk products from other species may expose us to signals designed to influence growth and development that are different from our evolved, adaptive, maternal milk signals. This is an area of current research. At present the data are sparse and the evidence indirect. But it is worth some speculation as to the extent that the signaling ability of milk, which we have outlined in the previous chapters of this book, might affect our physiology and metabolism when we consume another species' milk. Milk may be a food that comes closest in truth to the old saying, "you are what you eat."

Dairy Products in Antiquity

There is archeological evidence of the consumption of dairy products by modern humans dating back many thousands of years. People in Africa, the Middle East, South Asia, North-Central Europe, and Britain and Scotland were keeping domesticated ruminants (generally cows, sheep, and goats but also camels and horses) and using their milk for food. In some cases the raw milk was eaten; in many other instances, it was processed milk in some fashion, producing fermented milks and cheese and yogurt-like foods.

Milk fats, in particular long-chain fatty acids, can become absorbed into unglazed pottery and will withstand leaching and time (short-chain fatty acids are more water soluble and degrade faster). The stable carbon isotope signature of milk is distinguishable from fat from animal tissue. The stable carbon isotope signature of residues in ancient pottery can identify if rumi-nant milk products were contained within the pottery (Copley et al., 2003). This technique has been used to demonstrate the existence of some form of dairy industry extending back into the Neolithic in Britain, as long ago as 4000 BC (Copley et al., 2003; 2005a; 2005b; 2005c), in Switzerland from be-fore 3000 BC (Spangenberg et al., 2006), in Central Europe before 5500 BC (Craig et al., 2005), and in the Near East and southeastern Europe back to 7000 BC (Evershed et al., 2008).

Lactose, Lactase, and Human Evolution

The primary sugar in the milks of most species is the disaccharide lactose. This is certainly true for cow, goat, and sheep milk. The enzyme lactase cleaves the lactose molecule into its two constituent sugars, glucose and galactose. Lactase is required to digest lactose. In most humans and all other mammals after weaning and before adulthood the cells in their intes-tinal tract stop expressing the gene for lactase. Lactose intolerance, the inability to digest lactose, is the norm among adult mammals. From an evo-lutionary perspective, producing an enzyme for which there will be no sub-strate would be inefficient. Evolution does not always select for the most efficient phenotype. But lactase expression is a simple case of a peptide with no function post weaning; at least until humans began domesticating other animals.

There exists a technological solution, of course. Many bacteria will fer-ment lactose. Fermented milks, cheese, yogurt, and other milk products contain much less lactose, if any. Much of the dairy consumption long ago, at least by adult humans, was probably these kind of milk products. They can provide fat and high-quality protein, as well as calcium and phospho-rus, without providing the intestinal distress that comes from the fermenta-tion of undigested lactose in the large intestine. And of course raw milk itself would have been a good food for infants and young children who still produced lactase in their small intestine.

Some human populations have evolved lactase persistence; the continued expression of the lactase gene into adulthood. Several groups of humans, descended from herding cultures where milk became an important food resource, have a high incidence of mutations that inactivate the regulatory switch that turns off lactase expression. In these people lactase continues to be produced after childhood, and they thus remain lactose tolerant past childhood and for much if not all of their adult life. In the past this ability to digest milk appears to have provided a significant adaptive advantage.

Europeans tend to have a high incidence of lactase persistence. There is a gradient in the prevalence of lactase persistence, with lower levels in southern Europe and higher incidence as you travel north. In northernmost Europe (Great Britain, the Scandinavian countries) the mutation for lactase persistence is almost universal. The mutation that confers lactase persistence in European populations is thought to be a single nucleotide polymorphism (SNP) of the lactase gene (Tishkoff et al., 2007). But certain pastoral populations in Africa also exhibit significant levels of lactase persistence among adults, which appears to derive from three independent SNPs (Tishkoff et al., 2007; Ranciaro et al., 2014). Thus, several human populations appear to have independently produced adaptive mutations to the lactase gene that result in lactase persistence and the ability to digest lactose throughout life.

Strange Proteins

Milk from domesticated animals provided people with a dependable, high-quality food resource. But milk is not an unmixed blessing. Although milk from other species may be good food, it will contain proteins that differ from human milk and that could be potential allergens. For example, human milk lacks the milk protein β-lactoglobulin, which is found in all ruminant milks. β-lactoglobulin is one of the more common causes of milk allergies in people.

Cow milk can contain different versions of the β-casein molecule. The A2 variant of β-casein in cow milk is the ancestral form. A mutation in the β-casein gene that changes a single amino acid results in the A1 form. These β-casein variants have no effect on cow milk before it is ingested, but after ingestion the digestive process produces different breakdown products for these β-caseins variants. When digested by human peptidases, a 7-amino acid fragment of the A1 form of β-casein (called beta-casomorphin-7 or BCM-7) with opioid-like activity is created. BCM-7 has been shown to activate opiate receptors in the intestinal tract and to slow gastrointestinal passage rate through an opiate receptor pathway in rats (Barnett et al., 2014). There is preliminary evidence of an effect on gastrointestinal function in humans (Ho et al., 2014). More concerning, human populations that drink predominantly milk containing the A1 β-casein variant also show greater incidence of type 1 diabetes and heart disease (Laugesen and Elliott, 2003). Of course that is an association and does not prove causation, and the finding has been questioned by some researchers who note inconsistent study results and potential confounding parameters (e.g., Truswell, 2005; Merriman, 2009). But it has led to some level of controversy in the dairy industry, and the creation of A2-variant herds to produce A2-only milk by some producers.

It has also led to unsubstantiated speculation that milk and other dairy products, especially cheese, are addictive. This is largely based on the release of casomorphins from β-casein after digestion. These molecules (there are more than just BCM-7) can bind to and potentially activate opiate receptors. As noted above, BCM-7 does slow down gastrointestinal passage, an effect blocked by naloxone, an opiate receptor antagonist (Barnett et al., 2014). But this does not mean that casomorphins have potent opioid-like effects in brain. Injections of casomorphins had no addictive effects on adult rats; rats injected with casomorphin behaved similarly to placebo-injected rats, and different from the rats injected with morphine who did exhibit addiction-like symptoms (Reid and Hubbell, 1994). More importantly, casomorphins are unlikely to be able to pass healthy adult intestinal epithelium intact, and thus are unlikely to enter the adult bloodstream. There is no scientific evidence for an opioid-addiction-like effect of casomorphins.

The case may be different for neonates with immature intestinal tracts and adults with increased permeability of the digestive tract, where casomorphins may enter circulation and might have biological effects in periphery and brain. There is an association between autism and higher urinary casomorphins. This is generally assumed to be related to a higher intestinal permeability in many autistic children that allows a larger proportion of intact proteins to be absorbed into circulation. Gastrointestinal problems are associated with autism (van De Sande et al., 2014). In theory there might be a link between increased absorption of casomorphins from either cow milk or from breast milk (human β-casein also produces casomorphins upon digestion) consistent with the opioid excess theory of autism. This is the theory behind the gluten- and casomorphin-free diet to reduce food-related effects on autistic children. There is no conclusive direct evidence to support this theory, however (van De Sande et al., 2014).

So do casomorphins have biological effects (good or bad) in people? Does the A1 β-casein variant in cow milk contribute to disease in vulnerable populations? The evidence is uncertain, in our opinion. There are enough data for interesting scientific speculation, and to support further research in this area. An assessment by the European Food Safety Authority (2009) arrived at similar conclusions. The report found strong evidence that BCM-7 is a product of the A1 β-casein variant in cow milk, but little consistent evidence linking BCM-7 to disease in humans. Although BCM-7 is an opioid receptor ligand it appears to not be very potent in its effects. The absorption of intact BCM-7 (or other casomorphins) by healthy adults is questionable, but neonates and people with intestinal disorders that increase intestinal permeability may absorb them. Casomorphins are most likely to have effects on the intestinal tract, but may have broader biological effects in individuals who absorb them into circulation.

Signaling Molecules

Of course cow, goat, sheep, and other domesticated animal milk drunk by humans also contain the many biologically active signaling molecules we have discussed in this book: growth factors, metabolic hormones, cytokines, immune function molecules, and even non-coding RNA molecules. The concentrations and form of all these molecules in each species' milk was guided by its evolutionary history, initially, and then by human artificial selection, which was completely ignorant of their very existence and was focused on enhancing other, desired properties such as milk production or fat content. Pasteurization and homogenization of milk deactivate some, but not all, of these bioactive molecules. Some researchers have suggested that bioactive molecules in cow milk in particular have significant biological effects on children and possibly adults who consume it (e.g., Wiley, 2012). As you can imagine, the claims and counterclaims have political, commercial, and ideological elements to them. The dairy industry emphasizes potential benefits and downplays potential harms, while an anti-milk part of society does the opposite. The actual evidence for effects of signaling molecules in store-bought milk affecting the physiology of humans who consume it is still sparse, but not to be discounted.

As mentioned in chapter 10, milk contains vesicles, such as exosomes, that contain bioactive molecules including mRNA and miRNA. These exosomes are also capable of expressing cytokines from their membrane. Intriguing studies have indicated that these milk exosomes and their contents may have significant biological effects in people who drink milk. Exosomes from commercial cow milk were shown to express active transforming growth factor beta, and these exosomes were also shown to be taken up by mouse macrophages in vitro and could affect T-cell differentiation (Pieters et al., 2015). Cow milk consumption by the pregnant mother is associated with greater maternal weight gain, higher birth weight with an increase in fetal growth in the third trimester, and increased placenta weight (Melnik et al., 2015). There is substantial evidence that milk consumption in children enhances growth, both linear and body mass (e.g., Hoppe et al., 2006; Wiley, 2012).

Possible signaling mechanisms in cow milk suggested to underlie the association with enhanced growth include the fatty acid palmitate, branched-chain amino acids (e.g., leucine, isoleucine, and valine), and miRNAs. All these signaling molecules could potentially activate mTORC1, either directly or indirectly. mTORC1 is important in cell growth and proliferation as it regulates protein synthesis (Melnik et al., 2013; 2015). The potential activation of mTORC1 by miRNA-21, an miRNA commonly found in milk, was discussed in chapter 10.

Preliminary research indicates that miRNAs from cow milk can be absorbed into circulation after people consume a glass of milk, and that these miRNAs can be taken up by cells and affect gene expression. Consumption of milk by human subjects increased the concentration in plasma and in peripheral blood mononuclear cells (PBMCs) of two miRNAs (miRNA-29b and miRNA-200c) that are common in cow milk and which share 100% sequence identity between cow and human forms. Gene expression in PBMCs changed after milk consumption consistent with the predicted effects of these two miRNAs (Baier et al., 2014). These findings are fascinating indications that milk has the potential to exert regulatory effects even on adults. The study was also small and needs replicating with a greater diversity of miRNAs to be studied.

Summary

As we have evolved as a species after our divergence from the African great ape lineage, our milk has changed with us. It appears to have increased in fat, oligosaccharides, immune function molecules, metabolic hormones and growth factors, and the starch-digesting enzyme amylase. These changes are consistent with the growth of our large brains, the extra pathogen loads our ancestors suffered from before modern sanitation, hygiene, and medicine, and an early introduction of supplementary foods for our babies. Over the last approximately 10,000 years, the milk of other species—cow, goat, sheep, horse, camel, and so forth—has had a significant impact on our success as a species and, for the lactase-persistent populations, our genetic makeup. The use of other species' milk may still be influencing our growth patterns from in utero through childhood, and may affect our vulnerability, both positively and negatively, to certain diseases.

Chapter 13

Breastfeeding, History, and Health

Throughout history, philosophers and scientists have extolled the potency and importance of breast milk to the health and well-being of a human infant. In Book XII of *Attic Nights* (*Noctes Atticae*), the Roman author Aulus Gellius (ca. 125–180 AD) describes an incident in which the philosopher Favorinus (ca. 80–160 AD) broadly opines on the importance of a woman breastfeeding her baby. His words provide a fascinating window into the pseudoscientific knowledge of the time regarding mother's milk and infant development.

> the philosopher, having embraced and congratulated the father immediately upon entering, sat down. And when he had asked how long the labor had been and how difficult . . . he began to talk at greater length and said: "I have no doubt she will suckle her son herself!" But when the young woman's mother said to him that she must spare her daughter and provide nurses for the child, in order that to the pains which she had suffered in childbirth they might not be added the wearisome and difficult task of nursing, he said: "I beg you, madam, let her be wholly and entirely the mother of her own child. . . . To have nourished in her womb with her own blood something which she could not see, and not to feed with her own milk what she sees, now alive, now human, now calling for a mother's care? Or do you too perhaps think," said he, "that nature gave women nipples as a kind of beauty-spot, not for the purpose of nourishing their children, but as an adornment of their breast? . . .
>
> "'But it makes no difference,' for so they say, 'provided it be nourished and live, by whose milk that is effected.' Why then does not he who affirms this, . . . think that it also makes no difference in whose body and from whose blood a human being is formed and fashioned? Is the blood which is now in the breasts not the same that it was in the womb, merely because it has become white from abundant air and width? . . . just as the power and nature of the seed are able to form likenesses of body and mind, so the qualities and

properties of the milk have the same effect. And this is observed not only in human beings, but in beasts also; for if kids are fed on the milk of ewes, and lambs on that of goats, it is a fact that as a rule the wool is harsher in the former and the hair softer in the latter . . . What the mischief, then, is the reason for corrupting the nobility of body and mind of a newly born human being, formed from gifted seeds, by the alien and degenerate nourishment of another's milk? Especially if she whom you employ to furnish the milk is . . . dishonest, ugly, unchaste and a wine-bibber. . . ."

Modern science casts strong doubt on most of Favorinus's assertions. Cross-species drinking of milk will not affect the characteristics of a neonate's hair nor are moral failings (or virtues) transferred via milk. And yet the basic idea is true: important biological information that can profoundly affect infant development and the eventual health and well-being of the child is passed from mother to baby via milk. When a mother breastfeeds her baby she is engaging in one of the most intimate interactions with another person. She is feeding her baby with the product of her own body. But she is doing more than giving her child nutrition. She is transferring a wealth of bioactive molecules that will influence her baby's immune system, developmental patterns, and eventual adult physiology.

It has only been in the last few hundred years that a brief interlude of human arrogance questioned the primary importance of breast milk and suggested that humans could devise something better. That unfortunate episode of history perhaps can be encapsulated by two famous lines from Alexander Pope's 1709 *An Essay on Criticism*: "A little learning is a dangerous thing" and "Fools rush in where angels fear to tread."

Breastfeeding performs psychosocial, immunological, metabolic, and other regulatory functions in addition to its fundamental important nutritional role. All these milk-related factors interact to guide the growth and development of the baby. Breast milk provides much more than just food. Not surprisingly, breastfeeding and breast milk are associated with many aspects of health and disease, both during the neonatal period as well as later in life.

Human breast milk evolved to support the specific patterns of growth and development for our infants and also in response to immunological, sociological, and technological selective pressures in our past, both relatively recent (e.g., thousands or tens of thousands of years ago) and long ago (hundreds of thousands or millions of years ago). Human beings do not live like great apes. It is difficult to argue against the idea that the growth and development of our infants has changed more from that of infants of our common ancestor with the great apes than has the development of great ape infants. Our infants today are certainly exposed to many experiences and challenges that are completely novel compared to our distant and even not-so-distant past. At the same time, there are commonalities to the growth and development of all mammalian young. Human milk certainly evolved; but it is still milk, and it retains the basic ingredients to support milk's intrinsic function—producing independent offspring.

The biochemical complexity of milk, with its numerous and as-yet not completely characterized suite of bioactive molecules that could have developmental effects on babies, strongly suggests that there will be no complete substitutes for human milk anytime soon. Technology continues to advance, but millions of years of evolution have given human breast milk a substantial advantage over any formula humans might devise as a substitute. The hubris of early-twentieth-century scientists who believed they could do better than evolution has been shown to be nothing more than hubris.

That is not to say that breast milk is superior in all ways to any alternative or even the best option in all cases. Milk is an evolved substance, not a perfect substance. There are circumstances where breastfeeding is contraindicated, either due to maternal disease or to some rare but serious genetic conditions in the infant. Breast milk is also not required, as the many people who were formula-fed as babies and are fully functional, capable human beings demonstrate. However, epidemiological research suggests strongly that breastfed babies do accrue certain tangible and measurable benefits. What is lacking is decisive research linking specific factors in milk with many of these benefits, and proximate developmental and physiological mechanisms by which milk improves infant outcomes. We believe that such evidence will come sooner rather than later, now that our biological knowledge and our technical ability to investigate the complexity of milk have progressed. Some are already well understood, such as the benefits of milk immunoglobulins for resistance to disease. But for other benefits of breast milk the evidence is indirect and comes from epidemiological studies. We believe that breast milk confers an advantage, but we don't know exactly which molecules and mechanisms are involved. And the benefit may be transient, providing an advantage to babies but not necessarily carrying over into adulthood. For one thing, the parameters that affect health and well-being throughout life are complex. Being breastfed may provide an initial boost, but many factors later in life have profound effects as well. We believe that increasing the proportion of babies that are breastfed and the median duration of breastfeeding will have a positive public health benefit, but it certainly is not a silver bullet that inoculates an individual against bad outcomes.

In this chapter we present and evaluate some of the evidence for the tangible benefits of breastfeeding for infants, mothers, and society as a whole. The focus will be on evidence, not on theory. Theory we have already covered in the previous chapters. This chapter cannot be comprehensive; that would require a book of its own, and new evidence continues to be generated. We restrict ourselves to topics and examples that we find particularly interesting or instructive. We start with a brief history of human infant feeding.

Feeding Babies

For almost all of history, breast milk has been considered the best food for infants. It wasn't until the 1900s that the notion that foods other than breast milk might be better for babies gained wide acceptance. That is not to say that there isn't a long history of feeding babies foods other than breast milk; but the phrase "breast is best" would have resonated throughout history.

Not all perceptions of breast milk from ancient history survive scientific scrutiny. For much of history, the knowledgeable medical/scientific opinion was that colostrum was "bad" milk. Preserved writings from ancient Greek, Roman, and Indian Brahminical medicine recommended a mother should wait several days before suckling her child. Colostrum was taboo (Wickes, 1953). This prejudice against colostrum exists today in some cultures. The belief was not universal, and several influential writers recommended feeding infants colostrum. In the late 1700s, Hugh Smith wrote a strong endorsement of breastfeeding including giving colostrum to the baby. He recommended frequent suckling to stimulate lactation (Smith, 1772). Dr. Smith's views could be considered quite modern.

One difficulty in tracking the history of infant feeding practices is that writings about infants and children, especially scientific ones, are rare from before the 1800s. Most of the written works we have from our early historic past concern adults. Paulus Bagellardus of Padua wrote the first book of pediatrics of which we are aware, titled *A Little Book on the Diseases of Children*, published in 1472. In his book Bagellardus comments that mothers suckled their children for 2–3 years. In 1697, John Peachy

Table 13.1. A summary of recommendations by leading writers of the time for weaning age in the 1800s

1802	Wean at 6 to 8 months, provided four teeth are cut
1825	Wean at 8 to 10 months for a vigorous child, later if puny
1826	Wean at 11 to 12 months, but not unless several teeth are cut
1827	Wean after four incisors have been cut
1840	Wean at 9 months on average
1842	Wean at 7 to 12 months
1847	Wean according to the teeth. When canines appear more food needed
1853	Wean at 9 months, but remember teeth have meaning
1868	Wean at 12 months in ordinary cases
1886	Wean at 8 months
1897	Wean at 9 to 12 months

Compiled by Forsythe, 1911.

published "General Treatise of the Diseases of Infants and Children" in which he counsels that a child should suckle for 1.5–2 years and should not be weaned until the child has all its teeth. The duration of suggested breastfeeding continued to shorten over time, with the advice becoming when the child had most of his teeth in the 1700s to weaning when the first teeth in each jaw had been cut in the early 1800s (Forsythe, 1911). In the 1800s, the recommended weaning age varied between 6 and 12 months, sometimes referring to teeth as milestones, sometimes to other aspects of the infant's development (table 13.1). Over a relatively short stretch of history, the length of time a woman was counseled to breastfeed her infant (or have a wet nurse breastfeed it for her) had declined from as long as 3 years to under 1 year. The decline has continued; most women in the United States stop breastfeeding before their baby is six months old, and many babies never receive any breast milk.

Early weaning is a modern invention in part allowed by technology, especially refrigeration, and by artificial feeding by milk from other species, paps and gruels produced from breads and grains, and, nowadays, sophisticated manufactured milk replacer formulas and baby cereals. Human technological progress has reduced the importance of breast milk for modern babies. Women have more choices now for how to feed their babies: exclusive breastfeeding, breast milk via both breast and bottle, mixture of breastfeeding, formula, and other supplemental foods, or no breast milk at all. All these options can be successfully employed. From a public health perspective, if more women were to exclusively breastfeed for longer durations there would be a net benefit in the overall health of the population. We document the evidence behind that conclusion later in this chapter. But a woman's choice must account for many different aspects of her life and her baby's life, and while "breast is best" is true for most mother-baby dyads, reality is more complex and nuanced.

Wet Nursing

For most of our evolutionary history our babies were all breastfed. Even in the case of maternal death, if a neonate was to be kept alive it would have required another lactating women to accept and breastfeed it. How often that happened in our prehistory we will never know, but the earliest historical records indicate wet nurses have existed for at least 4,000 years. When the daughter of the Pharaoh found Moses, she employed a wet nurse (who was actually his own mother) to feed him. Evidence of occasional lactation failure is evident from the Egyptian medical encyclopedia *The Papyrus Ebers* dating from 1550 BC (Wickes, 1953). Wet nursing, the suckling of an infant other than one's own, is known from the beginning of history, and likely extends far back into our evolutionary past.

Wet nursing was an accepted profession from ancient Egypt through the Roman Empire and up until the late 1800s. Regulations regarding wet nurses survive from Hammurabi's Babylon (ca. 1700 BC). Written contracts for wet nurses date back to around 300 AD in the Roman Empire (Stevens et al., 2009). The emperor Nero was reportedly wet nursed (Wickes, 1953), as were probably most babies of the elites at that time in Rome. Between 100 and 400 AD, medical writers such as Soranus, Galen, and Oribasius provided lists of qualifications and advice for wet nurses (Stevens et al., 2009). These writings were the basis for most of the beliefs during the early Renaissance, with few documents from the Middle Ages about infant feeding practices surviving, if they existed at all (Wickes, 1953).

Most of the historical records support the idea that expert opinion considered an infant best served if nursed by its own mother. The main argument was that characteristics of the nurse were passed to the baby via milk. The words of Favorinus, as recorded by Aullus Geillus in his book *Attic Nights*, that start this chapter, encapsulate the general opinion regarding the danger of a mother not nursing her own children. Favorinus also warned that the affection of mother for the child and child for the mother could be adversely affected:

> For when the child is given to another and removed from its mother's sight, the strength of maternal ardor is gradually and little by little extinguished,

every call of impatient anxiety is silenced, and a child which has been given over to another to nurse is almost as completely forgotten as if it had been lost by death. Moreover, the child's own feelings of affection, fondness, and intimacy are centered wholly in the one by whom it is nursed, and therefore, just as happens in the case of those who are exposed at birth, it has no feeling for the mother who bore it and no regret for her loss.

These ideas did not stop the practice of wet nursing. Pliny the Elder and Tacitus bemoaned the disinclination of wealthy Roman women to breast-feed their babies, employing a wet nurse instead—and worried about the possible decline in civic values due to the supposed inheritance of failings from the wet nurse. Moralists during the Renaissance expressed similar concerns as wet nursing increased in popularity, drawing parallels between the decay of the Roman Empire and the practice of wet nursing for the elites (Forsythe, 1911).

Around 1600, the French obstetrician Jacques Guillemeau wrote a treatise translated into English as "The Nursing of Children." He provides four main objections to wet nurses: (1) that the baby may be switched with another; (2) that the natural affection between mother and child will decline; (3) that a bad condition or inclination will be derived from the wet nurse; and (4) that the wet nurse will transmit some bodily imperfection to the baby. The last concern appears to relate to disease, as Guillemeau worries that the baby may then transmit this imperfection to its parents.

The belief that characteristics of the wet nurse would be transmitted to the baby influenced the advice given on choosing a wet nurse, including such warnings as to avoid redheads, as their fiery temperament was detri-mental to the milk and the baby might inherit a treacherous mind (Forsythe, 1911). Islamic tradition encourages breastfeeding. If the mother is unable to breastfeed, a wet nurse is deemed acceptable, but not "adulteresses and the insane" whose milk is considered "infectious" (Shaikh and Ahmed, 2006). A belief in the transmission of characteristics through milk is ancient and widespread.

Despite these beliefs within the medical and scientific communities, the practice of wet nursing thrived through the 1800s and into the very early 1900s. The demand for wet nurses by wealthy woman (or their husbands) created a despicable situation where young, poor unmarried girls perceived an economic opportunity. They would become pregnant, send their baby out to a "baby farm," where it likely died, and hire on as a wet nurse to another infant. The demand for wet nurses increased the incidence of "still-born" babies among poor young women, as well (Forsythe, 1911). Laws were instituted eventually to regulate wet nurses and forbid such practices. For example, in France wet nurses were forbidden to nurse another baby until their own was 9 months old (Stevens et al., 2009).

What does modern science say? Milk certainly contains substances that affect infant development in multiple and profound ways, and the milk of mothers will vary in these constituents. We are pretty confident that the color of a woman's hair, eyes, or skin are not significant factors, but her disease and immunization history will be, and likely her metabolic state. In defense of wet nursing, the babies received human milk, with its evolved signals. Whether the baby of a wealthy woman who was nursed by a poorer woman benefited or not from the poor woman's disease history and metabolic state relative to the natural mother's is an interesting question, though likely never to be answered. The baby was better off than if it had been fed animal milk or other foods. Until modern sanitation, refrigeration, and an understanding of microbial contamination and resulting disease, babies fed something other than a woman's breast milk usually died (Forsythe, 1911). Employment of a wet nurse immediately after birth would have likely deprived the baby of colostrum, especially after laws to protect wet nurses (and especially their babies) was enacted in France. However, human milk continues to contain the important bioactive molecules throughout lactation. A wet-nursed baby would have been exposed to a lower concentration of these molecules, but not deprived. Also, the prejudice against colostrum handed down from the ancient Greeks and Romans meant that many babies were deprived of colostrum, even if nursed by their mothers.

What about the potential epigenetic signals in human milk, such as miR-NAs? Science has yet to provide a good answer. There is variation among women; but we don't know how significant it is, or the extent to which epigenetic signals in milk are transmitted to the infant. It is interesting that in many cultures, milk siblings—the child of and/or any child nursed by a wet nurse—are considered to have a kinship bond. The Quran forbids marriage between milk siblings. An intriguing hypothesis suggests that people nursed by the same woman may share epigenetic traits, providing a potential higher risk of transmitting a genetic/epigenetic disease to any children they might have together (Ozkan et al., 2012).

The practice of wet nursing exploded in the 1700s and 1800s, driven by economic and social factors (see Forsythe, 1911; Stevens et al., 2009 for reviews). The practice largely died due to the invention of the glass bottle, the rubber nipple, improved sterilization and sanitation methods in manufacturing of alternative foods, and refrigeration. These improvements dramatically reduced the transmission of disease from contamination of milk substitutes and were more convenient and cheaper than hiring a wet nurse.

Artificial Feeding

At some point in our evolutionary history, providing infants with foods other than breast milk became a part of successful human culture. Certainly this occurred by the time other animal species were domesticated, providing access to non-human milk. There are ancient Egyptian vessels that show evidence of having contained milk (that were also in a shape conducive to providing that milk to a baby) (Stevens et al., 2009). The evidence from the duplication of the salivary amylase gene and its expression in the mammary gland discussed in chapter 12 suggests that our prehistoric ancestor's infants were fed on more than mother's milk, and this tradition continues in all modern human cultures.

Through much of the early Renaissance there is little mention of feeding babies with milk from domesticated species. Perhaps the words of Favorinus raised concerns. Bad enough to take the chance a baby will inherit bad characteristics from a wet nurse, but if a lamb nursing from a goat changes the lamb, what might happen to a human baby fed cow milk? More likely there was no ability to keep fresh milk to feed to babies, at least in the cities. Evidence from pottery from ancient Egyptian, Grecian, and Roman sites indicate that milk from domesticated species was likely fed to babies (Wickes, 1953). Wickes (1953) suggests that there are few surviving written instructions for giving animal milk to babies because little instruction was needed, as long as fresh milk was available. The sanitary conditions and the lack of refrigeration would have been a major block to using fresh animal milk for anyone who did not own the animal. Peasants in the country likely made use of cow, goat, sheep, horse, and ass milk.

In the second half of the 1700s animal milk seems to have become acceptable for use. Foundlings were often fed animal milk, sometimes straight from the animal. The Hospice des Enfants Malades in Paris kept asses from which their foundlings suckled (figure 13.1). Alphonse Le Roy established a system by which goats nursed the foundlings under his care. The goats were trained to seek out their "baby" and straddle the crib to allow the baby to

Figure 13.1. Foundling babies at the Hospice des Enfants Malades in Paris nursing directly from asses kept for the purpose. From Sadler 1909.

suckle. He believed that "There is in milk, besides the different nutritious principles, an invisible element, the element of life itself, a fugitive gas so volatile that it escapes as soon as milk is in contact with the air" (quoted in Drake, 1930). In his opinion babies could not successfully be fed on milk expressed from the mammary, either human or non-human. In some way he may have been correct, if instead of an invisible element escaping from milk we consider microbes entering the milk. With no refrigeration and the microbial loads no doubt prevalent in the environment, milk straight from the mammary was more sanitary.

Technological progress through the 1800s changed infant feeding behavior at the beginning of the 1900s. The glass bottle and rubber nipples improved hygiene and convenience for bottle feeding. The manufacturing of condensed milk in sealed, sterile cans provided a sanitary product that could be safely kept on the shelf. Finally, refrigeration allowed even fresh milk to be stored safely, both by producers, middle men, and consumers. Bottle feeding a baby the milk from a cow or goat became more convenient and cheaper than hiring a wet nurse.

The beginning of the 1900s was also a time when the science of nutrition was beginning to explode. Previously, the important nutrients were considered to be fat, sugar, protein, and minerals. In the early 1900s, vitamins were discovered. The diseases of scurvy and rickets, both of which affected babies, were shown to be prevented by vitamins C and D, respectively. Fear of scurvy delayed the use of boiled or evaporated milk, as heat treatment destroys vitamin C; but by the 1920s the incidence of infantile scurvy was dramatically reduced due to the advice of Dr. Julius Hess (who could be considered the father of American neonatology) to supplement infant feedings with fruit and vegetable juices (Fomon, 2001). The use of canned milk for feeding infants increased.

The nutrient content of milks from different domesticated animals as well as human milk had been investigated starting in the mid- to late 1800s. Cow milk was known to have higher protein content than human milk, and to be qualitatively different as well (cow milk has a much greater proportion of casein, and so has a higher curds-to-whey ratio). Cow milk was known to need to be diluted and have sugar added to more closely resemble human milk, at least in protein and sugar content, but that resulted in a milk replacer that was low in energy, due to the dilution of fat. Non-milk additives were used to increase the energy density, with many mathematical formulas devised to assist the pediatrician in preparing an artificial milk—which explains the origin of the term "formula" for a milk replacer (Barness, 1987). Formulas began to be manufactured, with added vitamins and, later, iron, in both powdered and liquid form. These formulas replaced home recipes using evaporated cow milk. Diseases such as scurvy and rickets were dramatically reduced in babies (Barness, 1987). Better sanitation and refrigeration reduced diarrheal disease. There was reason to consider infant formulas a beneficial invention. Most pediatricians still considered breast milk superior, but often added the caveat "for the first few months." This was an acknowledgement of the immunological and anti-infectious factors in milk. Barness (1987; p. 170) ends his brief review of the history of infant feeding with this statement: "At present, for the first few months of life, human milk is . . . the nonpareil of infant feeding. Increasing demands on the quality of human milk substitutes provide the stimulus for improvement of such substitutes. However, it seems unlikely that nutrition alone can achieve all the goals."

The recommended duration of breastfeeding was now further reduced from 7–9 months down to the first few months of life. The nutritional aspects of human milk were presumed to be well understood, and thus to be able to be duplicated artificially; but there was a developing understanding that milk provides more than nutrition. The importance of the non-nutritive factors in milk were mainly thought of as anti-infectious, and to be primarily important in the first few months of life. There were arguments published that evolution had not produced the best food for infants but rather the best compromise food between maternal constraints and infant requirements (Dugdale, 1986). Many believed that science could produce a nutritionally superior product that focused on optimal function for infants, and cost the mother only money.

We have sympathy for the view that milk may not be perfect, as discussed in chapter 5; evolution may not produce perfection as defined by human criteria. Evolution has balanced multiple aspects and functions of milk to derive a fluid that has successfully produced humans for tens if not hundreds of thousands of years. However, we hope that the previous chapters have demonstrated to the reader that this perception of our ability to mimic all the functions of human milk with an artificial formula was overly optimistic. It was not unintelligent, but it was lacking in knowledge of the biochemical complexity of milk and of the processes of life. Science had only begun to understand how DNA is regulated, and the power of the various epigenetic mechanisms that have evolved. The peptide revolution was just beginning. We argue that we now know enough to realize that we don't know anywhere near enough to duplicate milk.

Have we managed to produce a satisfactory milk replacer formula? Based on the criteria of the 1800s the answer would be yes. Deaths among formula-fed babies are low. Formula-fed babies are likely to grow up to live normal, healthy lives. Our expectations have also increased, however. Breastfed babies still appear to have measureable advantages. Modern milk replacer formulas still lack important bioactive ingredients and cannot match the biological effects of breast milk.

Public Health

Over the last approximately 30 years since Barness wrote his review of infant feeding an accumulation of epidemiological evidence regarding the advantages of breastfeeding and breast milk over formula feeding has changed opinions of scientists, medical professionals, and public health policy decision-makers. The American Academy of Pediatrics (AAP) currently recommends exclusive breastfeeding for about the first 6 months of an infant's life with continued breastfeeding along with supplemented feeding for one year or longer (AAP, 2005). The American College of Obstetricians and Gynecologists (ACOG) agrees that exclusive breastfeeding should continue until the infant is approximately 6 months old and notes that continued breastfeeding after 6 months is beneficial (ACOG, 2007). The Healthy People 2010 initiative set the following goals for breastfeeding: 75% of all mothers to initiate breastfeeding, with 50% continuing for at least six months postpartum and 25% continuing to one year as well as 40% exclusively breastfeeding at 3 months and 17% exclusively breastfeeding at 6 months. The Healthy People 2020 goals have been raised to 81.9% of mothers initiating breastfeeding, with 61% continuing through at least 6 months and 34% breastfeeding at 1 year.

The advantages of breastfeeding over formula feeding for babies have been demonstrated in multiple studies. Breastfed infants are at lower risk for a number of childhood diseases, such as ear infections and diarrhea, and appear to be at decreased risk for developing obesity, asthma, and some allergies during infancy. A meta-analysis of studies regarding breastfeeding and the risk of sudden infant death syndrome (SIDS) found strong evidence for a dose-dependent protective effect of breastfeeding (Hauck et al., 2011). Any breastfeeding had a beneficial effect, but the effect was strongest for exclusive breastfeeding. Another meta-analysis showed a protective effect of breastfeeding against childhood leukemia. The results suggested that breastfeeding for 6 months or longer might decrease the incidence of leukemia by 14–19% (Amitay and Keinan-Boker, 2015). The advantages to the breastfed baby seem to be fairly well demonstrated.

What is less certain is the extent that any benefits carry over into later childhood and adulthood. The case of asthma is especially confusing. Well-designed studies tended to find a protective effect of breastfeeding for infants and very young children, but a possible increase in risk for asthma in older children, at least for those whose mothers' have asthma (e.g., Wright et al., 2001). Recent meta-analyses of the relationship between breastfeeding and later asthma have found a positive effect, but the quality of the data are low and the results very heterogenic (Dogaru et al., 2014; Lodge et al., 2015). The effect is strong for ages 0–2 years but becomes progressively reduced at older ages. In part, the heterogeneity of results reflects the heterogeneity of the diagnosis "asthma," especially in studies on young children. Many of the studies document "wheezing" episodes and equate those with asthma; but wheezing in babies and young children can be caused by viral illness. Breastfeeding has a demonstrated effect of reducing viral disease in babies, and thus the reduced "asthma" may be confounded with reduced viral infection (Kramer, 2014). At present, the evidence favors a benefit to breastfeeding in reducing "wheezing" episodes in children under 2, and a possible small reduction in the risk for actual asthma in older children.

One difficulty with evaluating the benefits of breastfeeding is that almost all the studies are by necessity observational. It is very difficult to design an ethical, randomized control trial. The closest was a large study conducted in Belarus by Dr. Michael Kramer in 1996–1997 in which 31 hospitals were randomly assigned either as controls or to have a breastfeeding support intervention. The intervention was successful, with a much higher proportion of mothers still breastfeeding at 3 months (43%) compared to mothers from the control hospitals (6%) (Kramer et al., 2001). This provided a randomly constructed set of babies who were mostly breastfed and a set who were not. The results were consistent with most studies of breastfeeding, in that breastfed babies had fewer bouts of diarrhea and less eczema; but follow-up studies 6.5 and 11.5 years later found no differences in disease risk, such as rates of obesity and asthma, blood pressure, or other symptoms of metabolic disease. The follow-up study at 6.5 years of age did find a significant increase in mean score on intelligence tests and teacher academic ratings among the breast-fed children (Kramer et al., 2008).

Breastfeeding and Obesity

Breastfed babies do not grow the same as formula-fed babies. The breastfed infant may grow faster in the first month of life, but generally slower over the first year. In a meta-analysis, formula-fed babies had higher lean mass at all measured time points (3–4 mo, 8–9 mo, 12 mo), consistent with higher weight gain in formula-fed infants. And fat mass was higher in formula-fed infants at 12 months (Gale et al., 2012). Infants at high risk for obesity (based on a set of risk factors) that were breastfed less than 2 months had an increased risk of their growth trajectory increasing over the first year. The longer a baby was breastfed the more likely it was to have a stable weight trajectory, and at-risk babies benefited most (Carling et al., 2015). Many people would value faster growth, especially in certain categories of infants such as preterm and small-for-gestational-age babies. But the evidence is accumulating, both in human studies and in animal models, that faster growth, including accelerated lean-mass growth, is strongly associated with higher adiposity and a vulnerability to obesity.

The protective effect of breastfeeding appears to be greatly attenuated after weaning. A large study that included a sibling cohort study, in which at least one sibling had been breastfed and another formula fed, questions the evidence of long-term protective effects of breastfeeding for obesity. In the sample as a whole, the percentage of obese children aged 4–14 years was greater for those who had been formula fed (17.38%) versus those who had been breastfed (11.91%). But among the discordant sibling sample (18.14% for formula fed versus 16.36% breastfed) the difference was no longer significant (Colen and Ramey, 2014). The effects from being siblings (genetics and environment) appear to outweigh the protective effect of breast milk, though it is interesting that the numerical difference was still in the same direction and that the breastfed siblings in the discordant sample were more likely to be obese than the full breastfed sample. Perhaps a factor that influences a vulnerability to obesity in families also influences the likelihood for breastfeeding?

A potential complicating factor is that an association between human milk oligosaccharides and body composition has been observed in breastfed babies (Alderete et al., 2015). A higher diversity of human milk oligosaccharides as well as higher concentrations of certain oligosaccharides were associated with lower body fat. Thus, even within breastfed babies there may be variation in body fat associated with signaling from mom.

Breastfeeding appears to reduce the incidence of fat babies and fat toddlers, which is a good thing. It may not have as strong an effect on the incidence of obese older children and adults. Not surprising, since so many other factors will influence a person's vulnerability to obesity.

Breastfeeding and Intelligence

The evidence for a positive effect of breastfeeding on intelligence (or at least on scores on assessments of intelligence) is relatively strong, though the size of the effect is not large. It takes large, well-designed studies to measure the effect. There have been many, however, and they generally return consistent results. In addition to the study by Kramer and colleagues (2008), which is the closest to a true randomized trial so far published, observational studies have found that, after controlling for potential confounding factors, breastfed babies do grow up to have higher intellectual and cognitive abilities, and a longer duration of breastfeeding results in a larger effect. This holds true for young adults (Mortensen et al., 2002) and young children (Belfort et al., 2013). But even these studies identify other parameters of importance. For example, the breastfeeding mother's fish consumption was positively associated with intellectual performance of her infant at three years of age (Belfort, 2013).

What We Confidently Know

On the plus side, breastfed babies are better protected against respiratory and gastrointestinal infections. They are at lower risk of SIDS. They tend to score slightly better in assessments of intelligence, at least in childhood, but possibly into early adulthood. While they are breastfeeding they are at lower risk for "wheezing" and eczema, although whether this protective effect extends into adulthood is uncertain. Their growth is probably more "normal," and they are less likely to accumulate excess adipose tissue.

On the negative side, they are more at risk of anemia and vitamin D deficiency, and the risk increases with the number of months they are exclusively breastfed. But those concerns are easily addressed with supplementation, such as with liquid vitamin D drops and the addition of iron-fortified cereals to the diet at the appropriate age.

What is the appropriate age for the addition of supplementary foods? Expert opinion varies. The recommendations from AAP and ACOG imply that significant supplementary foods, such as cereals, should not be added to the diet until about 6 months of age. The "about" reflects diversity of opinion and some ambiguity in the data. An age range of 4–6 months is often given as a good estimate, to allow for differences in maternal and infant circumstances and concerns. Indeed, recent studies have indicated that 3–7 months of age is a critical time period (probably related to intestinal immune system maturation) during which the introduction of cereals has the lowest chance of increasing the risk of bad outcomes, such as later celiac disease and diabetes (Norris et al., 2005; Ludvigsson and Fasano, 2012; Størdal et al., 2013). These risks are not great; but the evidence suggests that exclusive breastfeeding beyond 6 months of age provides risks for these diseases in addition to increased risk of anemia and vitamin D deficiency. The infant's intestinal tract appears to have a "sweet spot" in age for learning to handle novel proteins of after 3 months but before 6–7 months (Callahan, 2015).

Breastfeeding Is Good for Mom

There are also several health benefits of breastfeeding for the mother. Breastfeeding facilitates weight loss and enhances postpartum recovery by reducing postpartum bleeding. It also lowers a mother's risk of diabetes, breast and ovarian cancers, osteoporosis, and hip fracture (AAP, 2012). The release of oxytocin and prolactin by the suckling stimulus facilitates emotional bonding between mother and child. A decrease in depressive symptoms has also been associated with breastfeeding (Uvnas-Moberg, 1998; Thome et al., 2006), although causality can be tricky for this effect. Does breastfeeding reduce depression or does depression interfere with breastfeeding? A small study found suggestive evidence that breastfeeding may reduce a mother's later risk of Alzheimer's disease (Fox et al., 2013). Again, most of these studies are purely observational in nature and do not provide the quality of evidence a randomized, controlled trial would produce. Still, for most women there appear to be mostly good effects from breastfeeding and few biological downsides. The challenges to breastfeeding women come more from society and economics (e.g., lost worktime, potential effects on career path).

The interaction between breastfeeding and maternal obesity is especially important in our modern age. Breastfeeding does contribute to postpartum weight loss. But maternal obesity is a risk factor for poor lactation performance and lactation failure (Rasmussen, 2007). Obese women have a diminished prolactin response to nipple stimulation by suckling in the immediate postpartum period (Rasmussen and Kjolhede, 2004). The hormonal millieu of the obese breastfeeding woman differs from the evolved norm.

Breastfeeding also has societal, environmental, and economical advantages. Breastfed infants experience fewer illnesses and visits to a doctor, lowering direct medical expenses and also reducing indirect costs such as parent absenteeism from work. Formula is not cheap; breastfeeding lowers direct costs for families and for public assistance programs for mothers and children. Formula feeding has associated ecological and environmental challenges (e.g., disposal of bottles, cans, and liners) (Stenchever et al., 2007). Breastfeeding provides a public health benefit at little or no direct cost.

Not All Good in Breast Milk

Of course not all things transmitted to an infant via milk are beneficial to the neonate. Some pathogens can also be transmitted. Two recent case studies concluded likely transmission of yellow fever live virus from maternal vaccination to her infant, probably via breast milk (CDC, 2010a; Kuhn et al., 2011). West Nile virus also appears able to be passed through breast milk (CDC, 2002), though the incidence of such occurrences is rare (Hinckley et al., 2007). Of most importance to public health considerations, about 10% of maternal-child transmission of the HIV virus occurs via breast milk (Dunn et al., 1992; Nduati et al., 2000). This leads to a public health dilemma in areas of the developing world with both high HIV infection rates and poor access to safe water. Breastfeeding is an important factor protecting against diarrhea and other diseases. To avoid breastfeeding for fear of transmission of HIV may expose infants to increased risk of morbidity and mortality from water-borne diseases.

The efficiency of HIV viral transfer via breast milk is low. Without antiviral drugs about 10–15% of breastfed infants will contract HIV (Dunn et al., 1992; Coutsoudis et al., 2004). However, infants who are exclusively breastfed are at lower risk of acquiring HIV than are infants who are fed a mix of formula and breast milk or breast milk and supplemental foods (Coovadia et al., 2007). There appear to be factors in milk that inhibit HIV transfer. For example, HIV-infected women with above-average amounts of oligosaccharides in their milk were less likely to transmit the virus, especially if the oligosaccharides were also low in 3'-sialyllactose (Bode et al., 2012). With highly active antiretroviral therapy (HAART) less than 3% of breastfed infants will contract HIV. In a study of 102 HIV-infected mothers undergoing HAART in Uganda, no infants were diagnosed as HIV positive and the risk of infant death was six-fold higher among infants breastfed for less than 6 months (Homsy et al., 2010). In developing countries, breastfeeding in conjunction with HAART may be preferable to formula feeding despite the non-zero risk of HIV transmission.

There are novel environmental contaminants in breast milk due to pollution and the by-products of the modern industrial world that mothers absorb through ingesting contaminated water, food, or even through the air they breathe. Many of these man-made compounds have endocrine action or act as endocrine disruptors (e.g., PCBs, dioxins). The concentrations and diversity of these contaminants varies markedly between countries and within regions within a country. For example, breast milk samples from Denmark and Finland could be accurately distinguished by the relative concentrations of different dioxins (Krysiak-Baltyn et al., 2009). These bioactive milk contaminants have the potential to affect development. For example, phthalates are chemicals known to alter Leydig cell differentiation and function in rodent models, and are now ubiquitous in the modern environment. They are found in breast milk with wide variations in concentrations among different regions. The concentration of different phthalates in breast milk was associated with alterations in reproductive hormones in three-month-old infants in Denmark and Finland, with higher sex-hormone binding globulin, luteinizing hormone, and lower free testosterone in infants of mothers with high breast-milk phthalate concentrations (Main et al., 2006).

There are also naturally occurring toxins in the environment that can be transferred to neonates via breast milk. Domoic acid is a biotoxin produced by diatoms in the genus *Pseudo-nitzschia* that acts as a neurotoxin. Controlled studies in rodent models indicate that neonates are more susceptible than adults to this neurotoxin (Xi et al., 1997). During diatom blooms many filter-feeding shellfish can become contaminated with domoic acid, leading to poisoning of birds, marine mammals, and humans that feed on these shellfish. Lactating rats injected intraperitoneally with non-lethal doses of domoic acid rapidly cleared the toxin, but it did enter the milk at low levels, and was cleared from milk much slower than from blood. The maximum concentration of domoic acid in blood was 16 times higher than in milk, but after 8 hours the concentration in milk was 4 times higher than in blood (Maucher and Ramsdell, 2005). Although domoic acid was not detected in pups fed their mother's milk, it was detected in pups fed milk spiked with domoic acid to be roughly 10 times higher in concentration. However, even at this higher dose the amount absorbed by the pups was well below the concentration that causes pathology. It is likely that the doses of domoic acid required to poison a neonate through milk would be acutely toxic to the mother. This example serves both to warn that toxins and harmful chemicals can be transmitted through milk, but at the same time to caution that the concentrations often are very low and unlikely to affect the baby. Breastfeeding women should be careful about exposure to potential toxins, but it will usually not be a reason to cease breastfeeding.

Who Doesn't Breastfeed

The proportion of mothers breastfeeding in the United States is disappointing (Merz et al., 2002; AAP, 2005; Ahluwalia et al., 2005; Queenan, 2011). The only national objective for Healthy People 2010 that was met was that 75% of new mothers initiated breastfeeding. National rates of breastfeeding at 6 and 12 months and rates of exclusive breastfeeding at 3 and 6 months fall well below public health recommendations (CDC, 2010b). Since the release of the more ambitious Healthy People 2020 objectives, there is an even wider gap between breastfeeding outcomes and CDC goals (Chapman and Pérez-Escamilla, 2012).

Mothers with lower rates of breastfeeding tend to be young, lower-income, unmarried, less-educated, participants of the Supplemental Nutrition Program for Women, Infants, and Children (WIC), overweight or obese before pregnancy, and more likely to report their pregnancy was unintended (Forste et al., 2001; Ahluwalia et al., 2005; Hawkins et al., 2013; Jones et al., 2015). There are numerous barriers to breastfeeding that affect all women, such as mothers' complaints of pain/discomfort, embarrassment, employment, and inconvenience. However, there are also barriers that are specific to some minority women and/or reported more frequently among certain minority groups. Some barriers to breastfeeding that negatively affect low-income, minority women include: lack of social and cultural acceptance; language and literacy barriers; lack of maternal access to information that promotes and supports breastfeeding; acculturation; and lifestyle choices including smoking and alcohol use (Forste et al., 2001; Hurst, 2007; Nommsen-Rivers et al., 2010; Ringel-Kulka et al. 2011; Ramos, 2012). Furthermore, one of the largest studies that has examined the impact of nativity/ immigrant status, race/ethnicity, and socioeconomic factors on breastfeeding in the United States found that immigrant women in each racial/ethnic group had significantly higher rates of breastfeeding initiation and duration than native women (Singh et al., 2007). Immigrant children with foreign-born parents had the highest likelihood of being breastfed at 6 and 12 months, whereas the chances of not being breastfed at 6 and 12 months were twice as high among native children with native parents.

The history of breastfeeding for African American women is complex and may contribute to current negative attitudes. During the time of slavery, African American women were forced to serve as wet nurses for their master's children. Often they had to nurture and nurse their slave owner's children while their own children were left unattended or nursed by another mother (Blum, 1999). Historically, breastfeeding was considered a norm among all women regardless of race/ethnicity, socioeconomic status, education, or age (Stevens et al., 2009). However, when commercial formula companies began marketing artificial infant formula food as superior to natural breast milk, perceptions and practices of breastfeeding dramatically shifted. The use of infant formula was initially viewed as a symbol of status and wealth and moderate- to low-income families, particularly African Americans, were forced to continue breastfeeding or use evaporated milk products (Stevens et al., 2009). In 1988, the formula industry began marketing directly to the public and started to target middle- to lower-class families (Stevens et al., 2009). In order to better market their products, these companies began giving out free formula samples to hospitals and WIC, a federal nutritional organization primarily used by low-income, minority women (USDA, 2009). The combination of a negative perception of breastfeeding in the African American community, a high rate of poverty among African American mothers, and the offer of free infant formula to low-income women from WIC all likely contribute to the low breastfeeding rates in this community.

Several studies have found that WIC participation is strongly associated with low rates of initiating breastfeeding and early breastfeeding discontinuation particularly among African American and Hispanic women (Flower et al., 2008; Sparks, 2011). The CDC reported that among low-income women participating in WIC, the breastfeeding rate is 67.5% whereas for higher income WIC-eligible participants, the rate is 84.6% (CDC, 2010b). Beal, Kuhlthau, and Perrin (2003) found that African American women were less likely than white women to report having received breastfeeding advice from WIC counselors and more likely to report having received bottle-feeding advice from WIC counselors.

Hospitals in the United States are responding to the call from the CDC and others to institute policies and procedures to encourage breastfeeding. In 2013 more than half of US hospitals were using a majority of the CDC ten steps to increase breastfeeding. For nine out of ten steps the percentage of hospitals employing them increased between 2007 and 2013, with the only slight decrease in offering prenatal breastfeeding education, which was still at about 90%.

The Bottom Line

Breastfeeding is good for the baby and the mother and may impart long-term benefits to the baby that extend into adulthood. The evidence is strong for a few health conditions, especially for while the baby is still breastfeeding. The evidence for longer-term benefits is weaker, and in some cases not currently persuasive.

There are a few maternal and neonatal conditions that contraindicate breastfeeding, mostly rare, thankfully. The vast majority of mothers can choose to breastfeed their infant secure in the knowledge that they are doing the baby and themselves good. If more women in the United States breastfed their babies for longer there would be a public health benefit. There remain barriers to breastfeeding in the United States and around the world. In many cases these barriers are being reduced by concerted actions by governmental and non-governmental agencies, though many will argue there is still a long way to go. However, importantly, breastfeeding is not required. Formula-fed babies will grow up to be competent, fully functional adults.

All aspects of life present tradeoffs. One-size-fits-all recommendations rarely benefit everyone. Our hope is that women around the world are enabled to choose to breastfeed their babies if they decide that is the best option. Mothers (and their babies) deserve evidence-based recommendations grounded in good science. The evidence is sufficient to recommend breastfeeding as a beneficial choice, but not a necessary choice. More research on the effects of breastfeeding, especially on longer-term outcomes, would be useful.

Of course, as scientists, our bias is to call for and support more research. Science is a powerful way of knowing the world, and it is the viewpoint we are trained in and believe in. We caution that the science of milk is expanding rapidly, along with all biological science. Researchers are asking questions about milk that were unheard of not that long ago. In the 1900s scientists made the mistake of thinking they knew enough about milk to create substitutes. In some ways they were correct. The formulas were nutritionally satisfactory, but even the milk formula manufacturers now admit they were far from perfect and did not replicate all that is beneficial in milk. We expect research to continue to expand the list of bioactive properties of milk for some time to come. Milk is still an incompletely understood substance.

References

Aaby P, Marx C, Trautner S, et al. 2002. Thymus size at birth is associated with infant mortality: a community study from Guinea Bissau. Acta Paediatr 91: 698–703.

Aagaard K, Ma J, Antony KM, Ganu R, Petrosino J, Versalovic J. 2014. The placenta harbors a unique microbiome. Sci Transl Med 6:237ra65.

Abbondanza FN, Power ML, Dickson MA, Brown J, Oftedal OT. 2013. Variation in the composition of milk of Asian elephants (*Elephas maximus*) throughout lactation. Zoo Biol 32: 291–298.

Agarwal K, Mughal M, Upadhyay P, Berry J, Mawer E, Puliyel J. 2002. The impact of atmospheric pollution on vitamin D status of infants and toddlers in Delhi, India. Arch Dis Child 87: 111–113.

Ahluwalia I, Morrow B, Hsia J. 2005. Why do women stop breastfeeding? Findings from the Pregnancy Risk Assessment and Monitoring System. Pediatrics 116: 1408–1412.

Aiello LC, Wheeler P. 1995. The expensive-tissue hypothesis: the brain and the digestive system in human and primate evolution. Curr Anthropol 36: 199–221.

Alderete TL, Autran C, Breke BE, Knight R, Bode L, Goran MI, Fields DA. 2015. Associations between human milk oligosaccharides and infant body composition in the first 6 mo of life. Am J Clin Nutr doi: 10.3945/ajcn.115.115451.

Alfonso E, Rojas R, Herrera J, Ortega M, Lemus C, Cortez C, Ruiz J, Pinto R, Gomez H. 2012. Polymorphism of the prolactin gene (PRL) and its relationship with milk production in American Swiss Cattle. Afr J Biotechnol 11: 7338–7343.

Alsaweed, M, Hartmann, PE, Geddes, DT, Kakulas, F. 2015. MicroRNAs in breastmilk and the lactating breast: potential immunoprotectors and developmental regulators for the infant and the mother. Int J Environ Res Public Health 12: 13981–14020.

Alvarez JA, Ashraf A. 2010. Role of vitamin D in insulin secretion and insulin sensitivity for glucose homeostasis. Int J Endocrinol: 1–18, doi: 10.1155/2010/351385.

Alvarez-Erviti L, Seow Y, Yin H, Betts C, Lakhal S, Wood M. 2011. Delivery of siRNA to the mouse brain by systemic injection of targeted exosomes. Nat Biotechnol 29: 341–345.

American Academy of Pediatrics. 2005. Policy statement. Breastfeeding and the use of human milk. Pediatrics 115: 496–506.

American Academy of Pediatrics and American College of Obstetricians and Gynecologists. 2012. Guidelines to Perinatal Care, 7th ed.

American College of Obstetricians and Gynecologists. 2007. Committee Opinion No. 361. ObstetGynecol 109: 479–480.

Amitay E, Keinan-Boker L. 2015. Breastfeeding and childhood leukemia incidence: A meta-analysis and systematic review. JAMA Pediatr 169: e15025.

Anagnostou E, Sorrya L, Chaplin W, Bartz J, Halpern D, Wasserman S, Wang A, Pepa L, Tanel N, Kushki A, Hollander E. 2012. Intranasal oxytocin versus placebo in the treatment of adults with autism spectrum disorders: a randomized controlled trial. Mole Autism 3: 16.

Anderson JL, May HT, Horne BD, Bair TL, Hall NL, Carlquist JF, Lappé DL, Muhlestein JB, and Intermountain Heart Collaborative IHC Study Group. 2010. Relation of vitamin D deficiency to cardiovascular risk factors, disease status, and incident events in a general healthcare population. Am J Cardiol 106: 963–968.

Anderson J, McKinley K, Onugha J, Duazo P, Chernoff M, Quinn E. 2015. Lower levels of human milk adiponectin predict offspring weight for age: a study in a lean population of Filipinos. Matern Child Nutr, doi: 10.1111/mcn.12216.

Andersson Y, Sävman K, Bläckberg L, Hernell O. 2007. Pasteurization of mother's own milk reduces fat absorption and growth in preterm infants. Acta Paediatr 96: 1445–1449.

Anoop T, Jabbar P, Pappacahn J. 2010. Lactation associated with pituitary tumour in a man. CMAJ 182: 591.

Atiq M, Suria A, Nizami S, Ahmed I. 1998. Maternal vitamin-D deficiency in Pakistan. Acta Obstet Gynecol Scand 77: 970–973.

Attaie R, Richter R. 2000. Size distribution of fat globules in goat milk. J Dairy Sci 83: 940–944.

Ayers L, Missig G, Schulkin J, Rosen J. 2011. Oxytocin reduces background anxiety in a fear-potentiated startle paradigm: peripheral vs central administration. Neuropsychopharmacol 36: 2488–2497.

Azzarello M. 1991. Some questions concerning the Syngnathidae brood pouch. BMar Sci: 49741–49747.

Bai W, Yin R, Yang R, Khan W, Ma Z, Zhao S, Jiang W, Wang Z, Zhu Y, Luo G, et al. 2013. Technical note: identification of suitable normalizers for microRNA expresion analysis in milk somatic cells of the yak (*Bos grunniens*). J Dairy Sci 96: 4529–4534.

Baier S, Nguyen C, Xie F, Wood J, Zempleni J. 2014. MicroRNAs are absorbed in biologically meaningful amounts from nutritionally relevant doses of cow milk and affect gene expression in peripheral blood mononuclear cells, HEK-293 kidney cell cultures, and mouse livers. J Nutr 144: 1495–1500.

Baitchman E, Trusty M, Murphy H. 2007. Passive transfer of maternal antibodies to West Nile virus in flamingo chicks (*Phoenicopterus chilensis* and *Phoenicopterus ruber ruber*). J Zoo Wildl Med 38: 337–340.

Bakermans-Kranenburg MJ, van Ijzendoorn MH. 2013. Sniffing around oxytocin: review and meta-analyses of trials in healthy and clinical groups with implications for pharmacotherapy. Transl Psychiatry 3: e258.

Barker D. 1998. In utero programming of chronic disease. Clin Sci (Lond) 95: 115–128.

Barness SL. 1987. History of infant feeding practices. Am J Nutr 46: 168–170.

Barnett M, McNabb W, Roy N, Woodford K, Clarke A. 2014. Dietary A1 B-casein affects gastrointestinal transit time, diepitdyl peptidase-4 activity, and inflammatory status relative A2 B-casein in Wistar rats. Int J Food Sci Nutr 65: 720–727.

Bartol F, Wiley A, Bagnell C. 2008. Epigentic programming of porcine endometrial function and the lactocrine hypothesis. Reprod Domest Anim 43(Suppl. 2): 273–279.

Barr ML, Bertram EG. 1949. A morphological distinction between neurons of the male and female, and the behavior of the nucleolar satellite during accelerated nucleoprotein synthesis. Nature 163: 676–677.

Beal A, Kuhlthau K, & Perrin J. 2003. Breastfeeding advice given to African American and white women by physicians and WIC counselors. Public Health Rep 118: 368–376.

Beauchamp GK, Mennella JA. 2011. Flavor perception in human infants: development and functional significance. Digestion 83 Suppl 1: 1–6.

Belfort M. 2013. Infant feeding and childhood cognition at ages 3 and 7 years: effects of breastfeeding duration and exclusivity. JAMA Pediatr 167: 836–844.

Ben X, Chen R, Feng Z, Chen T, Zhang J. 2012. The benefits of expressed maternal milk and donor breast milk for preventing necrotizing enterocolitis in preterm infants: systematic review and meta-analysis. J Nutritional Dis Ther 2: 110.

Berglund SK, Westrup B, Hagglof B, Hernell O, Domellof M. 2013. Effects of iron supplmentation of LBW infants on cognition and behavior at 3 years. Pediatrics 131: 47–55.

Bernhart FW. 1961. Correlation between growth-rate of the suckling of various species and the percentage of total calories from protein in the milk. Nature 191: 358–360.

Bernstein R, Dominy N. 2013. Mount Pinatubo, inflammatory cytokines, and the immunological ecology of Aeta hunter-gatherers. Hum Biol 85: 231–250.

Berwaer M, Martial J, Davis J. 1994. Characterization of an upstream promoter directing extrapituitary expression of the human prolactin gene. Mol Endocrinol 8: 635–642.

Bezirtzoglou E, Tsiotsias A, Welling GW. 2011.Microbiota profile in feces of breast- and formula-fed newborns by using fluorescence in situ hybridization (FISH). Anaerobe 17: 478–482.

Bilko A, Altbacker V, Hudson R. 1994. Transmission of food preferences in the rabbit: the means of information transfer. Physiol Behav 56: 907–912.

Blanc B, Isliker H. 1961. Isolement et charactérization de la protéine rouge sidérophile du la matemel: La Iactotransferrine. BSoc Chim Fr Biol 43: 929.

Blum L. 1999. At the breast: Ideologies of breastfeeding and motherhood in the contemporary United States. Boston: Beacon Press.

Bode L. 2012. Human milk oligosaccharides: Every baby needs a sugar mama. Glycobiology 22: 1147–1162.

Bode L, Kulin L, Kim H-Y, Hsiao L, Nissan C, Sinkala M, Kankasa C, Mwiya M, Thea DM, Aldrovandi GM. 2012. Human milk oligosaccharide concentration and risk of postnatal transmission of HIV through breastfeeding. Am J Clin Nutr 96: 831–839.

Bolander F, Nicholas K, Van Wyk J, Topper Y. 1981. Insulin is essential for accumulation of casein mRNA in mouse mammary epithelial cells. Proc Natl Acad Sci 78: 5682–5684.

Bordet J, Mordet M. 1924. Le pouvoir bactériolytique du colosrum et du lait. Comp R Séaces Acad Sci 179: 1109–1113

Botha-Brink J, Modesto S. 2007. A mixed-age classed "pelycosaur" aggregation from South Africa: earliest evidence of parental care in amniotes. Proc R Soc B 274: 2829–2834.

Bouckenooghe T, Sisino G, Aurientis S, Chinetti-Gbaguidi G, Kerr-Conte J, Staels B, Fontaine P, Storme L, Pattou F, Vambergue A. 2014. Adipose tissue macrophages (ATM) of obese patients are releasing increased levels of prolactin during an inflammatory challenge: A role for prolactin in diabesity? Biochim BiophysActa 1842: 584–593.

Brainard G, Asch R, Reiter R. 1981. Circadian-rhythms of serum melatonin and prolactin in the rhesus-monkey (*Macaca mulatta*). Biomed Res-Tokyo 2: 291–297.

Brawand D, Wahli W, Kaessmann H. 2008. Loss of egg yolk genes in mammals and the origin of lactation and pleacentation. PLoS Biol 6: e63.

Bronsky J, Mitrova K, Karpisek M, Mazoch J, Durilova M, Fisarkova B, Stechova K, Prusa R, Nevoral J. 2011. Adiponectin, AFABP, and leptin in human breast milk during 12 months of lactation. J Pediatr Gastroenterol Nutr 52(4): 474–477.

Buckley J, Maunder RJ, Foey A, Pearce J, Val AL, Sloman KA. 2010. Biparental mucus feeding: a unique example of parental care in an Amazonian cichlid. J Exp Biol 213(Pt 22): 3787–3795.

Bullen J, Rogers H, Leigh L. 1972. Iron-binding proteins in milk and resistance to Escherichia coli infection in infants. Br Med J 1: 69–75.

Cabrera-Rubio R, Collado M, Laitinen K, Salminen S, Isolauri E, Mira A. 2012. Human milk microbiome changes over lactation and is shaped by maternal weight and mode of delivery. Am J Clin Nutr 96: 544–551.

Callahan A. 2015. The Science of Mom. Baltimore: Johns Hopkins University Press.

Calvo E, Galindo A, Aspres N. 1992. Iron status in exclusively breast-fed infants. Pediatrics 90: 375–379.

Canal-Macias M, Roncero-Martin R, Moran J, Lavado-Garcia J, Costa-Fernandez M, Pedrera-Zamorano J. 2013. Increased bone mineral density is associated with breastfeeding history in premenopausal Spanish women. Arch Med Sci 9: 703–708.

Carl G, Robbins C. 1988. The energetic cost of predator avoidance in neonatal ungulates: hiding versus following. Can J Zool 66: 239–246.

Carling S, Demment M, Kjohlede C, Olson C. 2015. Breastfeeding duration and weight gain trajectory in infancy. Pediatrics 135: 1392.

Carlini D, Stephan W. 2003. In vivo introduction of unpreferred synonymous codons into the drosophila Adh gene results in reduced levels of ADH protein. Genetics 163: 239–243.

Carroll R. 2009. The Rise of Amphibians: 365 Million Years of Evolution. Baltimore: Johns Hopkins University Press.

Cavanaugh J, Mustoe A, Taylor J, French J. 2014. Oxytocin facilitates fidelity in well-established marmoset pairs by reducing sociosexual behavior toward opposite-sex strangers. Psychoneuroendocrinol 49: 1–10.

Centers for Disease Control and Prevention, 2002. Possible West Nile virus transmission to an infant through breast-feeding: Michigan, 2002. MMWR 51, 877–878.

Centers for Disease Control and Prevention, 2010a. Transfusion-related transmission of yellow fever vaccine virus: California, 2009.MMWR. MMWR 59, 34–37.

Centers for Disease Control and Prevention. 2010b. Breastfeeding report card—United States, 2010. Retrieved from www.cdc.gov/breastfeeding/data/reportcard.htm.

Chapman D, Pérez-Escamilla R. 2012. Breastfeeding among minority women: Moving from risk factors to interventions 1–3. Adv Nutr 3: 95–104.

Charoenphandhu N, Wongdee K, Krishnamra N. 2010. Is prolactin the cardinal calciotropic maternal hormone? Trends Endocrinol Metab 21: 395–401.

Chen CC, Stairs D, Boxer R, Belka G, Horesman N, Alvarez J, Chodosh L. 2012. Autocrine prolactin induced by the Pten-Akt pathway is required for lactation initiation and provides a direct link between the Akt and Stat5 pathways. Gene Dev 26: 2154–2168.

Chen J, Frankshun A, Wiley A, Miller D, Welch K, Ho T, Bartol F, Bagnell C. 2011. Milkborne lactocrine-acting factors affect gene expression patterns in the developing neonatal porcine uterus. Reprod 141: 675–683.

Chen X, Gao C, Li H, Huang L, Sun Q, Dong Y, Tian C, Gao S, Dong H, Guan D, et al. 2010. Identification and characterization of microRNAs in raw milk during different periods of lactation, commercial fluid, and powdered milk products. Cell Res 20: 1128–1137.

Chipman D, Sharon N. 1969. Mechanism of lysozyme action. Science 165: 454–465.

Choleris E, Devidze N, Kavaliers M, Pfaff D. 2008. Steroidal/neuropeptide interactions in hypothalamus and amygdala related to social anxiety. Prog Brain Res 170: 291–303.

Choleris E, Gustfasson J, Korach K, Muglia L, Pfaff D, Ogawa S. 2003. An estrogen-dependent four-gene micronet regulating social recognition: a study with oxytocin and estrogen receptor-alpha and -beta knockout mice. Proc Natl Acad Sci USA 100: 6192–6197.

Choleris E, Ogawa S, Kavaliers M, Gustafsson JA, Korach KS, Muglia LJ, Pfaff DW. 2006. Involvement of estrogen receptor alpha, beta and oxytocin in social discrimination: A detailed behavioral analysis with knockout female mice. Genes Brain Behav 5: 528–539.

Christensen H, Murawsky M, Horseman N, Willson T, Gregerson K. 2013. Completely humanizing prolactin rescues infertility in prolactin knockout mice and leads to human prolactin expression in extrapituitary mouse tissues. Endocrinology 154: 4777–4789.

Clapp C, Weiner R. 1992. A specific, high affinity, saturable binding site for the 16-kilodalton fragment of prolactin on capillary endothelial cells. Endocrinology 130: 1380–1386.

Cnop M, Havel P, Utzschneider K, Carr D, Sinha M, Boyko E, Retzlaff B, Knopp R, Brunzell J, Kahn S. 2003. Relationship of adiponectin to body fat distribution, insulin sensitivity and plasma lipoproteins: evidence for independent roles of age and sex. Diabetologia 46: 459–469.

Coe C, Lubach G, Izard K. 1994. Progressive improvement in the transfer of maternal antibody across the Order Primates. Am J Primatol 32: 51–55.

Colen CG, Ramey DM. 2014. Is breast truly best? Estimating the effects of breastfeeding on long-term child health and well-being in the United States using sibling comparisons. Soc Sci Med 109: 55–65.

Collado M, Laitinen K, Salminen S, Isloauri E. 2012. Maternal weight and excessive weight gain during pregnancy modify the immunomodulatory potential of breast milk. Pediatr Res 72: 77–85.

Collinson A, Moore S, Cole T, Prentice A. 2003. Birth season and environmental influence on patterns of thymic growth in rural Gambian infants. Acta Paediatr 92: 1014–1020.

Cooke L, Fildes A. 2011. The impact of flavour exposure in utero and during milk feeding on food acceptance at weaning and beyond. Appetite 57(3): 808–811.

Coovadia HM, Rollins NC, Bland RM, Little K, Coutsoudis A, Bennish ML, Newell ML. 2007. Mother-to-child transmission of HIV-1 infection during exclusive breast-feeding in the first 6 months of life: an intervention cohort study. Lancet 369(9567): 1107–1116.

Copley MS, Berstan R, Dudd SN, Docherty G, Mukherjee AJ, Straker V, Payne S, Evershed RP. 2003. Direct chemical evidence for widespread dairying in prehistoric Britain. Proc Natl Acad Sci USA 100(4): 1524–1529.

Copley MS, Berstan R, Dudd SN, Straker V, Payne S, Evershed RP. 2005a. Dairying in antiquity. I. Evidence from absorbed lipid residues dating to the British Iron Age. J Archaeolog Sci 32(4): 485–503.

Copley MS, Berstan R, Mukherjee AJ, Dudd SN, Straker V, Payne S, Evershed RP. 2005b. Dairying in antiquity. III. Evidence from absorbed lipid residues dating to the British Neolithic. J Archaeolog Sc 32(4): 523–546.

Copley MS, Berstan R, Straker V, Payne S, Evershed RP. 2005c. Dairying in antiquity. II. Evidence from absorbed lipid residues dating to the British Bronze Age. J ArchaeologSci 32(4): 505–521.

Corner G. 1930. The hormonal control of lactation. Am J Physiol 95: 43–55.

Corthésy, B. 2007. Roundtrip ticket for secretory IgA: Role in mucosal homeostasis? J Immunol 178: 27–32.

Corthésy, B. 2013. Multi-faceted functions of secretory IgA at mucosal surfaces. Front Immunol 4, doi: 10.3389/fimmu.2013.00185.

Coursodon C, Dvorak B. 2012. Epidermal growth factor and necrotizing enterocolitis. Curr Opin Pediatr 24: 160–164.

Coutsoudis A, Dabis F, Fawzi W, Gaillard P, Haverkamp G, Harris DR, Jackson JB, Leroy V, Meda N, Msellati P, et al. 2004. Late postnatal transmission of HIV-1 in breast-fed children: an individual patient meta-analysis. J Infect Dis 189: 2154–2166.

Craig OE, Chapman J, Heron C, Willis LH, Bartosiewicz L, Taylor G, Whittle A, Collins M. 2005. Did the first farmers of central and eastern Europe produce dairy foods? Antiquity 79: 882–894.

Creel SR, Monfort SL, Wildt DE, Wasser PM. 1991. Spontaneous lactation is an adaptive result of pseudopregnancy. Nature 351: 660–662.

Creel SR, Waser PM. 1994. Inclusive fitness and reproductive strategies in dwarf mongooses. Behav Ecol 5: 339–348.

Croskerry P, Smith G, Leon M. 1978. Thermoregulation and the maternal behavior of the rat. Nature 273: 299–300.

Cunnane S, Crawford M. 2014. Energetic and nutritional constraints on infant brain development: implications for brain expansion during human evolution. J Hum Evol 77: 88–98.

Cunnane S, Stewart K. (eds.). 2010. Human brain evolution: the influence of freshwater and marine food resources. Hoboken: John Wiley & Sons.

Dadds M, MacDonald E, Cauchi A, Williams K, Levy F, Brennan J. 2014. Nasal oxytocin for social deficits in childhood autism: a randomized controlled trial. J Autism Dev Disord 44: 521–531.

Davidson L, Lönnerdal B. 1987. Persistence of human milk proteins in the breast-fed infant. Acta Paediatr Scand 76: 733–740.

Davies W. 1939. The composition of the crop milk of pigeons. Biochem J 33: 898–901.

Dawodu A, Tsang RC. 2012. Maternal vitamin D status: effect on milk vitamin D content and vitamin D status of breastfeeding infants. Adv Nutr 3(3): 353–361.

de Zegher F, Sabastiani G, Diaz M, Gómez-Roig M, López-Bermejo A, Ibáñez L. 2013. Breast-feeding vs formula-feeding for infants born small-for-gestational-age: divergent effects on fat mass and on circulating IGF-I and high-molecular-weight adiponectin in late infancy. J Clin Endocrinol Metab 98: 1242–1247.

de Zegher F, Sebastiani G, Diaz M, Sanchez-Infantes D, Lopez Bermejo A, Ibanez L. 2012. Body composition and circulating high-molecular-weight adiponectin and IGF-I in infants born small for gestational age: Breast- versus formula-feeding. Diabetes 61: 1969–73.

Diamond J. 1995. Father's Milk: From goats to people, males can be mammary mammals, too. Discover Mag. http://discovermagazine.com/1995/feb/fathersmilk468.

Dierenfeld E, Hintz H, Robertson J, Van Soest P, Oftedal O. 1982. Utilization of bamboo by the giant panda. J Nutr 112: 636–641.

Dixson A, George L. 1982: Prolactin and parental behavior in a male New World Primate. Nature 299: 551–553.

Dobbing J, Sands J. 1973. Quantitative growth and development of human brain. Arch Dis Child 48: 757–767.

Dogaru C, Nyffenegger D, Pescatore A, Spycher B, Kuehni C. 2014. Breastfeeding and childhood asthma: systematic review and meta-analysis. AmJ Epidemiol 179: 1153–1167.

Donovan SM, Wang M, Monaco MH, Martin CR, Davidson LA, Ivanov I, Chapkin RS. 2014. Noninvasive fingerprinting of host-microbiome interactions in neonates. FEBS Lett 588: 4112–4119.

Dotz V, Rudloff S, Meyer C, Lochnit G, Kunz C. 2015. Metabolic fate of neutral human milk oligosaccharides in exclusively breast-fed infants. Mol Nutr Food Res 59: 355–364.

Drake TGH. 1930. Infant feeding in England and in France from 1750 to 1800. Am J Dis Child 39: 1049–1061.

Dreu C. 2012. Oxytocin modulates cooperation within and completion between groups: An integrative review and research agenda. Horm Behav 61: 419–428.

Dugdale A. 1986. Evolution and infant feeding. Lancet 22: 670–673.

Dunn DT, Newell ML, Ades AE, Peckham CS. 1992. Risk of human immunodeficiency virus type 1 transmission through breastfeeding. Lancet 340: 585–588.

Duttaroy A. 2009. Transport of fatty acids across the human placenta: a review. Prog Lipid Res 48: 52–61.

Dvorak B, Fituch C, Williams C, Hurst N, Schnaler R. 2003. Increased epidermal growth factor levels in human milk of mothers with extremely premature infants. Pediatr Res54: 15–19.

Eberle M, Kappeler PM. 2006. Family insurance: kin selection and cooperative breeding in a solitary primate (*Microcebus murinus*). Behav Ecol Sociobiol 60: 582–588.

Ehrlich P, Hubner W. 1892. Uber die Ubertragung von Immunitat durch Milch [On the transfer of immunity by milk]. Deutsch Med Wochenschr, 18: 393.

Eisert R. 2011. Hypercarnivory and the brain: protein requirements of cats reconsidered. J Comp Physiol B 181: 1–17.

Eisert R, Oftedal O. 2009. Capital expenditure and income (foraging) during pinniped lactation: the example of the Weddell seal (*Leptonychotes weddellii*). In Smithsonian at the Poles: Contributions to International Polar Year Science, Krupnik I, Lang M, Miller S (eds.). Washington, DC: Smithsonian Institution Scholarly Press, pp. 335–346.

Eisert R, Oftedal O, and Barrell G. 2013. Milk composition in the Weddell seal (Leptonychotes weddellii): Evidence for a functional role of milk sugar in pinnipeds. Physiol Biochem Zool 86: 159–175.

Eising C, Eikenaar C, Schwabl H, Groothuis T. 2001. Maternal androgens in black-headed gull eggs. Proc R Soc Lond B 268: 839–846.

Eising C, Visser G, Muller W, Groothuis T. 2003. Steroids for free? No metabolic costs of elevated maternal androgen levels in the black-headed gull. J Exp Biol 206: 3211–3218.

Eldredge N, Gould S. 1972. Punctuated equilibria; an alternative to phyletic gradualism. In Models in Paleobiology, Schopf TJM (ed.). San Francisco: Freeman, Cooper and Co., pp. 82–115.

Ellis D, Whitlock P, Tsengeg P, Nelson R. 1999. Siblicide, splayed-toes-flight display and grappling in the Saker Falcon. J. Raptor Res 33: 164–167.

Else P. 2013. Dinosaur Lactation? J Exp Biol 216: 347–351.

Emera D, Casola C, Lynch V, Wildman D, Agnew D, Wagner G. 2012. Convergent evolution of endometrial prolactin expression in primates, mice, and elephants through the independent recruitment of transposable elements. Mol Biol Evol 29: 239–247.

European Food Safety Authority. 2009. Review of the potential health impact of β-casomorphins and related peptides. EFSA Scientific Report 231: 1–107.

Evershed RP, Payne S, Sherratt AG, Copley MS, Coolidge J, Urem-Kotsu D, Kostas Kotsakis MOG. 2008. Earliest date for milk use in the Near East and southeastern Europe linked to cattle herding. Nature 455: 528–531.

Falcon-Lang HJ, DiMichele WA, Elrick S, Nelson WJ. 2009. Going underground: in search of Carboniferous coal forests. Geology Today 25:181–184.

Febo M, Numan M, Ferris C. 2005. Functional magnetic resonance imaging shows oxytocin activates brain regions associated with mother-pup bonding during suckling. J Neurosci 25: 11637–11644.

Fernández L, Langa S, Martín V, Maldonado A, Jiménez E, Martín R, Rodríguez J. 2013. The human milk microbiota: origin and potential roles in health and disease. Pharmacol Res 69: 1–10.

Field C. 2005. The immunological components of human milk and their effect on immune development in infants. J Nutr 135: 1–4.

Fields D, Demerath E. 2012. Relationship of insulin, glucose, leptin, IL-6 and TNF-A in human breast milk with infant growth and body composition. Pediatr Obes 7: 304–312.

Flatt T, Heyland A. 2011. Mechanisms of life history evolution: the genetics and physiology of life history traits and trade-offs. Oxford University Press.

Flower K, Willoughby M, Cadigan R, Perrin E, Randolph G. 2008. Understanding breast-feeding initiation and continuation in rural communities: A combined qualitative/quan-titative approach. Mat and Child Health 12: 402–414.

Fomon S. 2001. Infant feeding in the 20th century: formula and beikost. J Nutr 131: 409–420S.

Forbes R, Cooper A, Mitchell H. 1953. The composition of the adult human body as deter-mined by chemical analysis. J Biol Chem 203: 359.

Forste R, Weiss J, Lippincott E. 2001. The decision to breastfeed in the United States: does race matter? Pediatrics 108: 291–296.

Forsythe D. 1911. The history of infant-feeding from Elizabethan Times. Proc R Soc Med 4: 110–141.

Fox M, Berzuini C, Knapp L. 2013. Maternal breastfeeding history and Alzheimer's disease risk. J Alzheim Dis: 1–13.

Frelinger J. 1971. Maternally derived transferrin in pigeon squabs. Science 171: 1260–1261.

Gale C, Logan C, Santhakumaran S, Parkinson J, Hyde M, Modi N. 2012. Effect of breast-feeding compared with formula feeding on infant body composition: a systematic review and meta-analysis. Am J Clin Nutr 95: 656–669.

Galef BG, Henderson PW. 1972. Mother's milk: a determinant of the feeding preferences of weaning rat pups. J Comp Physiol Psychol 78: 213–219.

Ganguli K, Meng D, Rautava S, Lu L, Walker WA, Nanthakumar N. 2013. Probiotics prevent necrotizing enterocolitis by modulating enterocyte genes that regulate immune-mediated inflammation. Am J Physiol Gastrointest Liver Physiol 304: G132–G141.

Garland CF, Garland FC. 1980. Do sunlight and vitamin D reduce the likelihood of colon cancer? Int J Epidemiol 9: 227–231.

Garly ML, Trautner SL, Marx C, Danebod K, Nielsen J, Ravn H, Martins CL, Balé C, Aaby P, Lisse IM. 2008. Thymus size at 6 months of age and subsequent child mortality. J Pediatr 153: 683–688.

Ghasemi N, Zadehrahmani M, Rahimi G, Hafezian S. 2009. Associations between prolactin gene polymorphism and milk production in montelbeliard cows. Int J Gen Mole Biol 3: 048–051.

Gibson RC, Buley KR, Douglas ME. 2004. Maternal care and obligatory oophagy in *Lepto-dactylus fallax*: a new reproductive mode in frogs. Copeia 1: 128–135.

Gillespie MJ, Stanley D, Chen H, Donald JA, Nicholas KR, Moore RJ, Crowley TM. 2012. Functional similarities between pigeon "milk" and mammalian milk: induction of im-mune gene expression and modification of the microbiota. PLoS One. 7:e48363.

Gluckman PD, Hanson MA. 2004. The developmental origins of the metabolic syndrome. Trends Endocrinol Metab 15: 183–187.

Gluckman PD, Hanson MA. 2006. Mismatch: Why Our World No Longer Fits Our Bodies. New York: Oxford University Press.

Gluckman PD, Lillycrop KA, Vickers MH, Pleasants AB, Phillips ES, Beedle AS, Burdge GC, Hanson MA. 2007. Metabolic plasticity during mammalian development is directionally dependent on early nutritional status. PNAS USA 104: 12796–12800.

Goldman AS. 2007. The immune system in human milk and the developing infant. Breast-feed Med 2: 195–204.

Goldman AS, Smith CW. 1973. Host resistance factors in human milk. J Pediat 82: 1082–1090.

Goldsmith A, Edwards C, Koprucu M, Silver R. 1981. Concentrations of prolactin and lu-teinizing hormone in plasma of doves in relation to incubation and development of the crop gland. J Endocrinol 90: 437–443.

Gonia S, Tuepker M, Heisel T, Autran C, Bode L, Gale CA. 2015. Human milk oligosaccha-rides inhibit *Candida albicans* invasion of human premature intestinal epithelial cells. J Nutr 145: 1992–1998.

Gordon C, Feldman H, Sinclair L, LeBoff W, Kleinman P, Perez-Rossello J, Cox J. 2008. Prevalence of vitamin D deficiency among healthy infants and toddlers. Arch Pediatr Adol Med 162: 505–512.

Gordon I, Vander Wyk B, Bennett R, Cordeaux C, Lucas M, Eilbott J, Zagoory-Sharon O, Leckman J, Feldman R, Pelphrey K. 2013. Oxytocin enhances brain function in children with autism. PNAS 110: 20953–20958.

Gordon J, Chitkara I, Wyon J. 1963. Weaning diarrhea. Am J Med Sci 245: 345.

Gorski J, Dunn-Meynell A, Hartman T, Levin B. 2006. Postnatal environment overrides genetic and prenatal factors influencing offspring obesity and insulin resistance. Am J Physiol—Regulatory, Integrative and Comparative Physiology 291: R768–R778.

Göth A, Booth D. 2005. Temperature-dependent sex ratio in a bird. Biol Lett 1: 31–33.

Goto K, Fukuda K, Senada A, Saito T, Kimura K, Glander K, Hinde K, Dittus W, Milligan L, Power M, et al. 2010. Chemical characterization of oligosaccharides in the milk of six species of new and old world monkeys. Glycoconjugate J 27: 703–715.

Goudswaard J, van der Donk JA, van der Gaag I, Noordzij A. 1979. Peculiar IgA transfer in the pigeon from mother to squab. Dev Comp Immunol 3(2): 307–319.

Gould L, Power ML, Elwanger N, Rambeloarivony H. 2011. Feeding behavior and nutrient intake in spiny forest-dwelling ring-tailed lemurs (*Lemur catta*) during early gestation and early to mid-lactation periods: Compensating in a harsh environment. Am J Phys Anthropol 145: 469–479.

Gould L, Sussman RW, Sauther ML. 2003. Demographic and life history patterns in a population of ring-tailed lemurs (*Lemur catta*) at Beza Mahafaly Reserve, Madagascar: a 15-year perspective. Am J Phys Anthropol 120: 182–194.

Gould S, Eldredge N. 1977. Punctuated equilibria: the tempo and mode of evolution reconsidered. Paleobiology 3: 115–151.

Grant W, Garland C. 2004. A critical review of studies on vitamin D in relation to colorectal cancer. Nutr Cancer 48: 115–123.

Green A, Ramsey C, Villaverde C, Asami D, Wei A, Fascetti A. 2008. Cats are able to adapt protein oxidation to protein intake provided their requirement for dietary protein is met. J Nutr 138: 1053–1060.

Grey K, Davis E, Sandman C, Glynn L. 2012. Human milk cortisol is associated with infant temperament. Psychoneuroendocrinol 38: 1178–1185.

Gritli-Linde A, Hallberg K, Harfe BD, Reyahi A, Kannius-Janson M, Nilsson J, Cobourne MT, Sharpe PT, McMahon AP, Linde A. 2007. Abnormal hair development and apparent follicular transformation to mammary gland in the absence of hedgehog signaling. Dev Cell 12: 99–112.

Grölund MM, Lehtonen OP, Eerola E, Kero P. 1999. Fecal microflora in healthy infants born by different methods of delivery: permanent changes in intestinal flora after cesarean delivery. J Pediatr Gastroenterol Nutr 28: 19–25.

Groves C, Grubb P. 2011. Ungulate Taxonomy. Baltimore: Johns Hopkins University Press.

Gu Y, Li M, Wang T, Liang Y, Zhong Z, Wang X, Zhou Q, Chen L, Lang Q, He Z, et al. 2012. Lactation-related microRNA expression profiles of porcine breast milk exosomes. PLoS ONE 7(8): e43691.

Guareschi C. 1936. Necessita di fattori alimentari materni per l'accrescimento del giovanissimi colombi. Boll Soc Ital Biol Sper 11: 411–412.

Gutkowska J, Jankowski M, Mukaddam-Daher S, McCan S. 2000. Oxytocin is a cardiovascular hormone. Braz J Med Biol Res 33: 625–633.

Hajjoubi S, Rival-Gervier S, Hayes H, Floriot S, Eggen A, Piumi F, Chardon P, Houdebine LM, Thépot D. 2006. Ruminants genome no longer contains Whey Acidic Protein gene but only a pseudogene. Gene 370: 104–112.

Hanson L. 1961. Comparative immunological studies of the immune globulins of human milk and blood serum. International Archives of Allergy and Applied Immunology 18: 241.

Harmon R, Schanbacher F, Ferguson L, Smith K. 1976. Changes in lactoferrin, immunoglobulin G, bovine serum albumin, and alpha-lactalbumin during acute experimental and natural coliform mastitis in cows. Infect Immun 13: 533–542.

Hata T, Murakami K, Nakatani H, Yamamoto Y, Matsuda T, Aoki N. 2010. Isolation of bovine milk-derived micro-vesicles carrying mRNAs and microRNAs. Biochem Biophy Res Co 396: 528–533.

Hauck F, Thopmson J, Tanabe K, Moon R, Vennemann M. 2011. Breastfeeding and reduced risk of sudden infant death syndrome: a meta-analysis. Pediatrics 128: 1–8.

Hausner H, Nicklaus S, Issanchou S, Molgaard C, Moller P. 2010. Breastfeeding facilitates acceptance of a novel dietary flavor compound. Clin Nutr 29: 141–148.

Hawkins SS, Stern AD, Gillman M W. 2013. Do state breastfeeding laws in the USA promote breast feeding? J Epidemiol Commun H 67: 250–256.

Hazewinkel HA, How KL, Bosch R, Goedegebuure SA, Voorhout G. 1987. Of vitamin D in dogs. In Nutrition, malnutrition and dietetics in the dog and cat: proceedings of an international symposium held in Hanover, September 3 to 4, 1987 (p. 66). British Veterinary Association in collaboration with the Waltham Centre for Pet Nutrition.

Healthy People 2010: Understanding and Improving Health. 2nd ed. US Department of Health and Human Services. Washington, DC: U.S. Government Printing Office, November 2000.

Healthy People 2020: Leading Health Indicators: Progress Update. US Department of Health and Human Services. Washington, DC: U.S. Government Printing Office, March 2014.

Hegardt P, Lindbert T, Borjesson J, Skude G. 1984. Amylase in human milk from mothers of preterm and term infants. J Pediatr Gastroenrerol Nutr 3: 563–566.

Heid H, Keenan T. 2005. Intracellular origin and secretion of milk fat globules. Eur J Cell Biol 84: 245–258.

Hennighausen L, Robinson G, Wagner K, Liu X. 1997. Prolactin signaling in mammary gland development. J Biol Chem272: 7567–7569.

Hepper P. 1995. Human fetal "olfactory" learning. Int J Prenatal Perinatal Psychol and Med 7: 147–151.

Hildeman W. 1959. A cichlid fish, *Symphysodon discus*, with unique nurture habits. Am Nat 93: 27–34.

Hillier LW, Miller W, Birney E, Warren W, Hardison RC, Ponting CP, Bork P, Burt DW, Groenen M, Delany ME, et al. 2004. Sequence and comparative analysis of the chicken genome provide unique perspectives on vertebrate evolution. Nature 432: 695–716.

Hillman L. 1990. Mineral and vitamin D adequacy in infants fed human milk or formula between 6 and 12 month of age. J Pediatr 117: S134–142.

Hinckley AF, O'Leary DR, Hayes EB. 2007. Transmission of West Nile virus through human breast milk seems to be rare. Pediatrics 119: e666–e671.

Hinde K. 2009. Richer milk for sons but more milk for daughters: sex-biased investment during lactation varies with maternal life history in rhesus macaques. Am J Hum Biol 21: 512–519.

Hinde K, Capitanio J. 2010. Lactational programming? mother's milk energy predicts infant behavior and temperament in rhesus macaques (*Macaca mulatta*). Am J Primatol 72: 522–529.

Hinde K, Carpenter AJ, Clay JS, Bradford BJ. 2014. Holsteins favor heifers, not bulls: biased milk production programmed during pregnancy as a function of fetal sex. PLoS ONE 9: e86169.

Hinde K, Foster AB, Landis LM, Rendina D, Oftedal OT, Power ML. 2013. Daughter dearest: sex-biased calcium concentrations in mother's milk among rhesus macaques. Am J Phys Anthropol 151: 144–150.

Hinde K, Skibiel A, Foster A, Del Rosso L, Menoza A, Capitanio J. 2015. Cortisol in mother's milk across lactation reflects maternal life history and predicts infant temperament. Behav Ecol 26: 269–281.

Ho S, Woodford K, Kukuljan S, Pal S. 2014. Comparative effects of A1 versus A2 beta-casein on gastrointestinal measures: a blinded randomised cross-over pilot study. Eur J Clin Nutr 68: 994–1000.

Holick, MF. 2003. Evolution and function of vitamin D. In Vitamin D Analogs in Cancer Prevention and Therapy, Reichrath J, Friedrich M, Tilgen W (eds.). Springer Berlin Heidelberg. pp. 3–28

Holt C. 1983. Swelling of golgi vesicles in mammary secretory-cells and its relation to the yield and quantitative composition of milk. J Theor Biol 101: 247–261.

Holt C, Carver J. 2012. Darwinian transformation of a "scarcely nutritious fluid" into milk. J Evolution Biol 25: 1253–1263.

Holt C, Carver J, Ecroyd H, Thorn D. 2013. Caseins and the casein micelle: their biological functions, structures and behavior in foods. J Dairy Sci 96: 6127–6146.

Homsy J, Moore D, Barasa A, Were W, Likichiho C, Waiswa B, Downing R, Malamba S, Tappero J, Mermin J. 2010. Breastfeeding, mother-to-child HIV transmission, and mortality among infants born to HIV-infected women on highly active antiretroviral therapy in rural Uganda. J Acq Immun Def Synd 53: 28–35.

Hoppe C, Mølgaard C, Michaelsen KF. 2006. Cow's milk and linear growth in industrialized and developing countries. Annu Rev Nutr 26: 131–173.

Horseman N, Gregerson K. 2014. Prolactin actions. J Mol Endocrinol 52: R95–106.

How KL, Hazewinkel HAW, Mol JA. 1994. Dietary vitamin D dependence of cat and dog due to inadequate cutaneous synthesis of vitamin D. Gen Comp Endocrinol 96: 12–18.

How KL, Hazewinkel HAW, Mol JA. 1995. Photosynthesis of vitamin D in the skin of dogs, cats, and rats. Vet Quart 17: S29.

Human Microbiome Project Consortium. 2012. Structure, function and diversity of the healthy human microbiome. Nature 486: 207–214.

Hurst C. 2007. Addressing breastfeeding disparities in social work. Health Soc Work 32: 207–210.

Irwin D. 1995. Evolution of the bovine lysozyme gene family: changes in gene expression and reversion of function. J Mol Evol 41: 299–312.

Irwin D, Biegel J, Stewart C. 2011. Evolution of mammalian lysozyme gene family. BMC Evol Biol 11: 166.

Isolauri E. 2012. Development of healthy gut microbiota early in life. J Paediatr Child H 48: 1–6.

Ito C, Shinkai A. 1993. Mother-young interactions during the brood-care period in *Anelosimus crassipes* (Araneae: Theridiidae). Acta Arachnologica 42: 73–81.

Izumi H, Kosaka N, Shimizu T, Sekine K, Ochiya T, Takase M. 2012. Bovine milk contains microRNA and messenger RNA that are stable under degradative conditions. J. Dairy Sci. 95: 4831–4841.

Izumi H, Kosaka N, Shimizu T, Sekine K, Ochiya T, Takase M. 2014. Time-dependent expression profiles of microRNAs and mRNAs in rat milk whey. PLoS ONE 9(2): e88843.

Jabed A, Wagner S, McCracken J, Wells D, Laible G. 2012. Targeted microRNA expression in dairy cattle directs production of B-lactoglublin-free, high-casein milk. Proc Natl Acad Sci USA 109: 16811–16816.

Jain V, Gupta N, Kalaivani M, Jain A, Sinha A, Agarwal R. 2011. Vitamin D deficiency in healthy breastfed term infants at 3 months and their mothers in India: seasonal variation and determinants. Indian J Med Res 133: 267–273.

Jankowski M, Hajjar F, Kawas S, Mukaddam-Daher S, Hoffman G, McCan S, Gutkowska J. 1998. Rat heart: a site of oxytocin production and action. Proc Natl Acad Sci USA 95: 14558–14563.

Jauniaux E, Gulbis B, Burton G. 2003. The human trimester gestational sac limits rather than facilitates oxygen transfer to the foetus—a review. Placenta 24A: 286–293.

Jenab M, Bueno de Mesquita H, Ferrari P, van Duijnhoven F, Norat T, Pischon T, Jansen E, Slimani N, Byrnes G, Rinaldi S, et al. 2010. Association between pre-diagnostic circulating vitamin D concentration and risk of colorectal cancer in European populations: a nested case-control study. Brit Med J 340: b5500.

Ji Q, Luo Z-X, Yuan C-X. Tabrum A. 2006. A swimming mammaliaform from the Middle Jurassic and ecomorphological diversification of early mammals. Science 311: 1123–1127.

Jones KM, Power ML, Queenan JT, Schulkin J. 2015. Racial and ethnic disparities in breast-feeding. Breastfeed Med 10: 186–196.

Kalkwarf H, Specker B, Bianchi D, Ranz J, Ho M. 1997. The effect of calcium supplementation on bone density during lactation and after weaning. New Engl J Med 337: 523–528.

Kato K, Ikemoto T, Park M. 2005. Identification of the reptilian prolactin and its receptor cDNAs in the leopard gecko, *Eublepharis macularius*. Gene 346: 267–276.

Kaufman S, Mackay B. 1983. Plasma prolactin levels and body fluid deficits in the rat: causal interactions and control of water intake. J Physiol 336: 73–81.

Kawakami H, Lönnerdal B. 1991. Isolation and function of a receptor for human lactoferrin in human fetal intestinal brush-border membranes. Am J Physiol 261: G841–846.

Kawasaki K, Lafont A, Sire J. 2011. The evolution of milk casein genes from tooth genes before the origin of mammals. Mol Biol Evol 28: 2053–2061.

Kennett JE, McKee DT. 2012. Oxytocin: an emerging regulator of prolactin secretion in the female rat. J Neuroendocrinol 24: 403–412.

Kenny DE, Irlbeck NA, Eller JL. 1999. Rickets in two hand-reared polar bear (*Ursus maritimus*) cubs. J Zoo Wildlife Med 30: 132–140.

Khong HK, Kuah MK, Jaya-Ram A, Shu-Chien AC. 2009. Prolactin receptor mRNA is upregulated in discus fish (*Symphysodon aequifasciata*) skin during parental phase. Comp Biochem Phys B Biochem Mol Biol 152: 18–28.

Kim KW, Roland C. 2000. Trophic egg laying in the spider, *Amaurobius ferox*: mother-offspring interactions and functional value. Behav Process 50: 31–42.

Kim KW, Roland C, Horel A. 2000. Functional value of matriphagy in the spider *Amaurobius ferox*. Ethology 106: 729–742.

Kim SH, Bennett P, Terzidou V. 2015. Diverse roles of oxytocin. Inflamm Cell Signal2: e739.

Kohler P, Farr R. 1966. Elevation of cord over maternal IgG immunoglobulin: evidence for an active placental IgG transport. Nature 210: 1070–1071.

Kokay I, Bull P, Davis R, Ludwig M, Grattan D. 2006. Expression of the long form of the prolactin receptor in magnocellular oxytocin neurons is associated with specific prolactin regulation of oxytocin neurons. Am J Physiol-Reg I 290: R1216–1225.

Kosaka N, Izumi H, Sekine K, Ochiya T. 2010. microRNA as a new immune-regulatory agent in breast milk. Silence 1: 1–8.

Kramer MS. 2014. Invited commentary: does breastfeeding protect against "asthma"? Am J Epidemiol 179: 1168–1170.

Kramer MS, Aboud F, Mironova E, Vanilovich I, Platt RW, Matush L, Igumnov S, Fombonne E, Bogdanovich N, Ducruet T, et al. 2008. Breastfeeding and child cognitive development: new evidence from a large randomized trial. Arch Gen Psychiat65: 578–584.

Kramer MS, Chalmers B, Hodnett ED, Sevkovskaya Z, Dzikovich I, Shapiro S, Collet JP, Vanilovich I, Mezen I, Ducruet T, et al. 2001. Promotion of Breastfeeeding Intervention Trial (PROBIT): a randomized trial in the Republic of Belarus. JAMA 285: 413–420.

Kramer K, Choe C, Carter C, Cushing B. 2006. Developmental effects of oxytocin on neural activation and neuropeptide release in response to social stimuli. Horm Behav 49: 206–214.

Kritz-Silverstein D, Barrett-Connor E, Hollenbach K. 1992. Pregnancy and lactation as determinants of bone mineral density in postmenopausal women. Am J Epidemiol 136: 1052–1059.

Krysiak-Baltyn K, Toppari J, Skakkebaek NE, Jensen TS, Virtanen HE, Schramm KW, Shen H, Vartiainen T, Kiviranta H, Taboureau O, et al. 2009. Country-specific chemical signatures of persistent environmental compounds in breast milk. Int J Androl 32: 1–9.

Kudo S, Nakahira T, Saito Y. 2006. Morphology of trophic eggs and ovarian dynamics in the subsocial bug *Adomerus triguttulus* (Heteroptera: Cydnidae). Can J Zool 84: 723–728.

Kuhn S, Twele-Montecinos L, MacDonald J, Webster P, Law B. 2011. Case report: probable transmission of vaccine strain of yellow fever virus to an infant via breast milk. CMAJ 183: E243–E245.

Kumar A, Rai AK, Basu S, Dash D, Singh JS. 2008. Cord blood and breast milk iron status in maternal anemia. Pediatrics 121(3): e673–677.

Kunz T, Hosken D. 2009. Male lactation: why, why not and is it care? Trends Ecol Evol 24: 80–85.

Kupfer A, Muller H, Jared C, Antoniazzi M, Nussbaum R, Greven H, Wilkinson M. 2006. Parental investment by skin feeding in a caecilian amphibian. Nature 440: 926–929.

Kupfer A, Wilkinson M, Gower D, Muller H, Jehle R. 2008. Care and parentage in a skin-feeding caecilian amphibian. J Exp Zool Part A 309A: 460–467.

Kuzawa C. 1998. Adipose tissue in human infancy and childhood: an evolutionary perspective. Am J Phys Anthropol 27: 177–209.

Langer, P. 2008. The phases of maternal investment in eutherian mammals. Zool 111:148–162.

Lash G, Robson S, Bulmer J. 2010. Review: Functional role of uterine natural killer (uNK) cells in human early pregnancy decidua. Placenta 24: S87–S92.

Laskey M, Prentice A. 1997. Effect of pregnancy on recovery of lactational bone loss. Lancet 349: 1518–1519.

Laugesen M, Elliott R. 2003. Ischaemic heart disease, type 1 diabetes, and cow milk A1 B-casein. New Zeal Med J 116: 1–19.

Lee A, Cool R, Grunwald W, Neal D, Buckmaster C, Cheng M, Hyde S, Lyons D, Parker K. 2011. A novel form of oxytocin in New World monkeys. Biol Lett 7: 584–587.

Lee RC, Feinbaum RL, Ambros V. 1993. The C. elegans heterochronic gene lin-4 encodes small RNAs with antisense complementarity to lin-14. Cell 75: 843–854.

Lefèvre CM, Sharp JA, Nicholas KR. 2009. Characterisation of monotreme caseins reveals lineage-specific expansion of an ancestral casein locus in mammals. Reprod Fert Develop 21: 1015–1027.

Lent P. 1974. Mother-infant relationships in ungulates. In The Behavior of Ungulates and Its Relation to Management, Geist V, Walther F (eds.). Morges, Switzerland: International Union for Conservation of Nature and Natural resources, pp. 147–157.

Leon M, Croskerry P, Smith G. 1978. Thermal control of mother-young contact in rats. Physiol Behav 21: 793–811.

Ley SH, Hanley AJ, Sermer M, Zinman B, O'Connor DL. 2012. Associations of prenatal metabolic abnormalities with insulin and adiponectin concentrations in human milk. Am J Clin Nutr 95: 867–874.

Li X, Li X, Huang HY, Ma D, Zhu MW, Lin JF. 2009. Correlations between adipocytokines and insulin resistance in women with polycystic ovary syndrome. Zhonghua Yi Xue Za Zhi 89(37): 2607–2610.

Libersat F, Gal R. 2014. Wasp voodoo rituals, venom cocktails, and the zombification of cockroach hosts. Integrative Comp Biol 54: 129–142.

Licht P, Bennett A. 1972. A scaleless snake: tests of the role of reptilian scales in water loss and heat transfer. Copeia: 702–707.

Lillywhite H. 2006. Water relations of tetrapod integument. J Exp Biol 209: 202–226.

Lincoln D, Renfree M. 1981. Mammary gland growth and milk ejection in the agile wallaby, *Macropus agilis*, displaying concurrent asynchronous lactation. J Reprod Fertil 63: 193–203.

Lindberg T, Skude G. 1982. Amylase in human milk. Pediatrics 70: 235–238.

Liu B, Newburg DS. 2013. Human milk glycoproteins protect infants against human pathogens. Breastfeed Med 8: 354–362.

Liu Y, Wang Z. 2003. Nucleus accumbens oxytocin and dopamine interact to regulate pair bond formation in female prairie voles. Neuroscience 121: 537–544.

Llewellyn D, Lang I, Langa K, Muniz-Terrera G, Phillips C, Cherubini A, Ferrucci L, Melzer D. 2010. Vitamin D and risk of cognitive decline in elderly persons. Arch Intern Med 170: 1135–1141.

Lodge CJ, Tan DJ, Lau M, Dai X, Tham R, Lowe AJ, Bowatte G, Allen KJ, Dharmage SC. 2015. Breastfeeding and asthma and allergies: a systematic review and meta-analysis. Acta Pediatr Suppl 104(467): 38–53.

Long J, Trinajstic K, Young G, Senden T. 2008. Live birth in the Devonian period. Nature 453: 650–652.

Ludvigsson J, Fasano A. 2012. Timing of introduction of gluten and celiac disease risk. Ann Nutr Metab 60: 22–29.

Luo ZX, Crompton AW, Sun AL. 2001. A new mammaliaform from the early Jurassic and evolution of mammalian characteristics. Science 292: 1535–1540.

Lydersen C, Kovacs KM. 1999. Behaviour and energetics of ice-breeding, North Atlantic phocid seals during the lactation period. Mar Ecol Prog-Ser 187: 265–281.

Macciò A, Madeddu C, Chessa P, Panzone F, Lissoni P, Mantovani G. 2010. Oxytocin both increases proliferative response of peripheral blood lymphomonocytes to phytohemagglutinin and reverses immunosuppressive estrogen activity. In Vivo 24: 157–163.

Macpherson AJ. Uhr T. 2004. Induction of protective IgA by intestinal dendritic cells carrying commensal bacteria. Science 303: 1662–1665.

Maguire J, Salehi L, Birken C, Carsley S, Mamdani M, Thrope K, Lebovic G, Khovratovich M, Parkin P, TARGet Kids! Collaboration. 2013. Association between total duration of breastfeeding and iron deficiency. Pediatrics 131: e1530–1537.

Maier A, Chabanet C, Schaal B, Leathwood P, Issanchou S. 2008. Breastfeeding and experience with variety early in weaning increase infants' acceptance of new foods for up to two months. Clin Nutr 27: 849–57.

Main KM, Mortensen GK, Kaleva MM, Boisen KA, Damgaard IN, ChellakootyM, Schmidt IM, Suomi A-M, Virtanen HE, Petersen JH, et al. 2006. Human breast milk contamination with phthalates and alterations of endogenous reproductive hormones in infants three months of age. EnvironHealth Persp 114: 270–276.

Malonza P, Measey G. 2005. Life history of an African caecilian: *Boulengerula taitanus* Loveridge 1935 (Amphibia Gymnophiona Caeciilidae). Trop Zool 18: 49–66.

Marano RJ, Ben-Jonathan N. 2014. Minireview: Extrapituitary prolactin: An update on the distribution, regulation, and functions. Mole Endocrinol 28: 622–633.

Martin R. 1966. Tree shrews: unique reproductive mechanism of systematic importance. Science 3: 1402–1404.

Martín R, Langa S, Reviriego C, Jiménez E, Marín ML, Olivares M, Boza J, Jiménez J, Fernández L, Xaus J, Rodríguez M. 2004. The commensal microflora of human milk: new perspectives for food bacteriotherapy and probiotics. Trends Food Sci Tech 15: 121–127.

Martín V, Maldonado-Barragán A, Moles L, Rodriguez-Baños M, del Campo R, Fernández L, Rodríguez JM, Jiménez E. 2012. Sharing bacterial strains between breast milk and infant feces. J Hum Lact 28: 36–44.

Masson P, Heremans J, Schonne E. 1969. Lactoferrin, an iron-binding protein in neutrophilic leukocytes. J Exp Med 130: 643–658.

Mata LJ, Wyatt RG. 1971. The uniqueness of human milk: host resistance to infection. Am J Clin Nutr 24: 976–986.

Mather I. 2000. A review and proposed nomenclature for major proteins of the milk-fat globule membrane. J Dairy Sci 83: 203–247.

Mather I, Keenan T. 1998. Origin and secretion of milk lipids. J Mammary Gland Biol 3: 259–273.

Matsuda A, Jacob A, Wu R, Zhou M, Nicastro J, Coppa G, Wang P. 2011. Milk fat globule-EGF factor VIII in sepsis and ischemia-reperfusion injury. Mol Med 17: 126–133.

Maucher J, Ramsdell J. 2005. Domoic acid transfer to milk: evaluation of a potential route of neonatal exposure. Environ Health Persp113: 461–464.

Mayer JA, Foley J, De La Cruz D, Chuong C-M, Widelitz R. 2008. Conversion of the nipple to hair-bearing epithelia by lowering bone morphogenetic protein pathway activity at the dermal-epidermal interface. Am J Pathol 173: 1339–1348.

McCollum E, Simmonds N, Shipley P, Park E. 1922. Studies on experimental rickets: XVI. A delicate biological test for calcium-depositing substances. Bull Johns Hopkins Hospital 33: 239.

Mellanby T. 1918. The part played by an "accessory factor" in the production of experimental rickets. J Physiol 52: 11–14.

Melnik BC, John SM, Schmitz G. 2015. Milk consumption during pregnancy increases birth weight, a risk factor for the development of diseases in civilization. 2015. J Transl Med 13: 13.

Melnik BC, John SM, Schmitz G. 2013. Milk is not just food but most likely a genetic transfection system activating mTORC1 signaling for postnatal growth. Nutr J 12: 103. PMCID: PMC3725179.

Melton III LJ, Bryant SC, Wahner HW, O'Fallon WM, Malkasian GD, Judd HL, Riggs BL. 1993. Influence of breastfeeding and other reproductive factors on bone mass later in life. Osteoporosis Int 3: 76–83.

Mennella JA, Beauchamp GK. 1991. Maternal diet alters the sensory qualities of human milk and the nursling's behavior. Pediatr 88: 737–744.

Mennella JA, Beauchamp GK. 1996. Developmental changes in the acceptance of protein hydrolysate formula. J Dev Behav Pediatr 17: 386–91.

Mennella JA, Beauchamp GK. 1998. Development and bad taste. Pediatr Asthma Allergy Immunol 12: 161–163.

Mennella JA, Blumberg MS, McClintock MK, Moltz H. 1990. Inter-litter competition and communal nursing among Norway rats: advantages of birth synchrony. Behav Ecol Sociobiol 27: 183–190.

Mennella JA, Forestell CA, Morgan LK, Beauchamp GK. 2009. Early milk feeding influences taste acceptance and liking during infancy. Am J Clin Nutr 90 (suppl): 780S–788S.

Mennella JA, Jagnow CP, Beauchamp GK. 2001. Prenatal and postnatal flavor learning by human infants. Pediatrics 107: E88.

Mennella JA, Johnson A, Beauchamp GK. 1995. Garlic ingestion by pregnant women alters the odor of amniotic fluid. Chem Senses 20: 207–209.

Mennella JA, Lukasewycz LD, Castor SM, Beauchamp GK. 2011. The timing and duration of a sensitive period in human flavor learning: a randomized trial. Am J Clin Nutr 93(5): 1019–1024.

Merewood A, Mehta S, Grossman X, Chen T, Mathieu J, Holick M, Bauchner H. 2012. Vitamin D status among 4-month old infants in New England: a prospective cohort study. J Hum Lact 28: 159–166.

Merriman TR. 2009. Type 1 diabetes, the A1 milk hypothesis and vitamin D deficiency. Diabetes Res Clin Pr, 83: 149–156.

Merz C, Buse J, Tuncer D, Twillman G. 2002. Survey on diabetes and heart disease physician attitudes and practices and patient awareness of the cardiovascular complications of diabetes. J Am Coll Cardiol 40: 1877–1881.

Milligan L. 2007, Dissertation. Nonhuman primate milk composition: relationship to phylogeny, ontogeny, and ecology. Tucson: University of Arizona.

Milligan L, Bazinet P. 2008. Evolutionary modifications of human milk composition: evidence from long-chain polyunsaturated fatty acid composition of anthropoid milks. J Hum Evol 55: 1086–1095.

Milligan LA, Gibson SV, Williams LE, Power ML. 2008a. The composition of milk from Bolivian Squirrel monkeys (*Saimiri boliviensis boliviensis*). Am J Primatol 70:35–43.

Milligan LA, Rapoport S, Cranfield M, Dittus W, Glander K, Oftedal O, Power M, Whittier C, Bazinet R. 2008b. Fatty acid composition of wild anthropoid priamtemilks. Comp Biochem PhysB 149: 74–82.

Missig G, Ayers L, Schulkin J, Rosen J. 2010. Oxytocin reduces background anxiety in a fear-potentiated startle paradigm. Neuropsychopharmacol 35: 2607–2616.

Mivart, S.G.J. 1871. On the Genesis of Species. London: Macmillan.

Moore S, Cole T, Collinson A, Poskitt E, McGregor I, Prentice A. 1999. Prenatal or early postnatal events predict infectious deaths in young adulthood in rural Africa. Int J Epidemiol28: 1088–1095.

Moore S, Fulford A, Wagatsuma Y, Persson L, Arifeen S, Prentice A. 2014. Thymus development and infant and child mortality in rural Bangladesh. Int J Epidemiol 43: 216–223.

Morange M. 2011. What history tells us XXIV. The attempt of Nikolai Koltzoff (Koltsov) to link genetics, embryology and physical chemistry. J Bioscience 32: 211–214.

Morris J, Earle K, Anderson P. 1999. Plasma 25-hydroxyvitamin D in growing kittens is related to dietary intake of cholecalciferol. J Nutr 129: 909–912.

Mortensen E, Michaelsen K, Sanders S, Reinisch J. 2002. The association between duration of breastfeeding and adult intelligence. JAMA 287: 2365–2372.

Mossberg A, Mok K, Morozova-Roche L, Svanborg C. 2010. Structure and function of human a-lactalbumin made lethal to tumor cells (HAMLET)-type complexes. FEBS J 277: 4614–4625.

Munch EM, Harris RA, Mohammad M, Benham AL, Pejerrey SM, Showalter L, Hu M, Shope, CD, Maningat, PD, Gunaratne, PH, et al. 2013. Transcriptome profiling of micro-RNA by next-gen deep sequencing reveals known and novel miRNA species in the lipid fraction of human breast milk. PLoS One 8(2):e50564. PMCID: PMC3572105.

Murphy W, Eizirik E, O'Brien S, Madsen O, Scally M, Douady C, Teeling E, Ryder O, Stanhope, M, de Jong W, et al. 2001. Resolution of the early placental mammal radiation using Bayesian phylogenetics. Science 294: 2348–2352.

Myburgh J, Osthoff G, Hugo A, de Wit M, Fourie D. 2012. Comparison of the milk composition of free-ranging indigenous African cattle breeds. S Afr J Wildl Res 42: 23–34.

Nathanielsz P. 2006. Animal models that elucidate basic principles of the developmental origins of adult diseases. Int League of Ass for Rheumatol J 47: 73–82.

Navara K, Hill G, Mendonca M. 2005. Variable effects of yolk androgens on growth, survival, and immunity in eastern bluebird nestlings. Physiol Biochem Zool 78: 570–578.

Nduati R, John G, Mbori-Ngacha D, Richardson B, Overbaugh J, Mwatha A, Ndinya-Achola J, Bwayo J, Onyango FE, Hughes J, et al. 2000. Effect of breastfeeding and formula feeding on transmission of HIV-1. JAMA 283: 1167–1174.

Nelson W, Pfiffner J. 1931. Studies on the physiology of lactation. I. The relation of lactation to the ovarian and hypophyseal hormones. Anat Rec 51: 51–83.

Neu J, Walker W. 2011. Necrotizing Enterocolitis. New Engl J Med 364: 255–264.

Newburg D, Peterson J, Ruiz-Palacios G, Matson D, Morrow A, Shults J, Guerrero M, Chaturvedi P, Newburg S, Scallan C, et al. 1998. Role of human-milk lactadherin in protection against symptomatic rotavirus infection. Lancet 351: 1160–1164.

Newburg D, Woo J, Morrow A. 2010. Characteristics and potential functions of human milk adiponectin. J Pediatr 156: S41–46.

Newman C, Cohen J, Kipnis C. 1985. Neodarwinian evolution implies punctuated equilibria. Nature 315: 400–401.

Ngom P, Collinson A, Pido-Lopez J, Henson S, Prentice A, Aspinall R. 2004. Improved thymus function in exclusively breastfed infants is associated with higher interleukin 7 concentrations in their mothers' breast milk. Am J Clin Nutr 80: 722–728.

Nicholas K. 1988. Control of milk protein synthesis in the marsupial *Macropus eugenii*: a model system to study prolactin-dependent development. In The Developing Marsupial. Models for Biomedical Research, Tyndale-Biscoe C, Janssens P (eds.). Berlin: Springer-Verlag, pp. 69–85.

Nie Y, Speakman J, Wu Q, Zhang C, Hu Y, Xia M, Yan L, Hambly C, Wang L, Wei W, et al. 2015. Exceptionally low daily energy expenditure in the bamboo-eating giant panda. Science 349: 171–174.

Nolte D, Provenza F. 1992. Food preferences in lambs after exposure to flavors in milk. Appl Anim Behav Sci 32: 381–389.

Nommsen-Rivers LA, Chantry CJ, Cohen RJ, Dewey KG. 2010. Comfort with the idea of formula feeding helps explain ethnic disparity in breastfeeding intentions among expectant first-time mothers. Breastfeed Med 5: 25–33.

Norris J, Barriga K, Hoffenberg E, Taki I, Miao D, Haas J, Emery L, Sokol R, Erlich H, Eisenbarth G, et al. 2005. Risk of celiac disease autoimmunity and timing of gluten introduction in the diet of infants at increased risk of disease. JAMA 293: 2343–2351.

Numan M, Fleming A, Levy F. 2006. Maternal behavior. In Knobil and Neill's Physiology of Reproduction, third ed. Neill JD (ed.), St. Louis: Elsevier, pp. 1921–1993.

Oben JA, Mouralidarane A, Samuelsson AM, Matthews PJ, Morgan ML, McKee C, Soeda J, Fernandez-Twinn DS, Martin-Gronert MS, Ozanne SE, et al. 2010. Maternal obesity during pregnancy and lactation programs the development of offspring non-alcoholic fatty liver disease in mice. J Hepatol 52: 913–920.

Oftedal, O. 2002a. The mammary gland and its origin during synapsid evolution. J Mammary Gland Biol 7: 225–252.

Oftedal, O. 2002b. The origin of lactation as a water source for parchment-shelled eggs. J Mammary Gland Biol 7: 253–266.

Oftedal, O. 2012. The evolution of milk secretion and its ancient origins. Animal 6: 355–368.

Oftedal, O. 2013. Origin and evolution of the major constituents of milk. In Advanced Dairy Chemistry: Volume 1A: Proteins: Basic Aspects, 4th edition. McSweeney PLH, Fox PF (eds.). New York: Springer, pp. 1–42.

Oftedal O, Alt G, Widdowson E, Jakubasz M.1993b. Nutrition and growth of suckling black bears (*Ursus americanus*) during their mothers' winter fast. Brit J Nutr 70: 59–79.

Oftedal O, Bowen WD, Boness DJ. 1993a. Energy transfer by lactating hooded seals and nutrient deposition in their pups during the four days from birth to weaning. Physiol Zool 66: 412–436.

Oftedal O, Iverson SJ. 1995. Comparative analysis of nonhuman milks: phylogenetic variation in the gross composition of milks. In: Handbook of Milk Composition, Jensen RG (ed). San Diego: Academic, pp 749–788.

Ogg SL, Weldon AK, Dobbie L, Smith AJ, Mather IH. 2004. Expression of butyrophilin (Btn1a1) in lactating mammary gland is essential for the regulated secretion of milk-lipid droplets. Proc Natl Acad Sci USA 101: 10084–10089.

Ohno S, Kaplan WD, Kinosita R. 1959. Formation of the sex chromatin by a single X-chromosome in liver cells of *Rattus norvegicus*. Exp Cell Res 18: 415–418.

Okyay D, Okyay E, Dogan E, Kurtulmus S, Acet F, Taner C. 2013. Prolonged breast-feeding is an independent risk factor for postmenopausal osteoporosis. Maturitas 74: 270–275.

Osei K. 2010. 25-OH vitamin D: is it the universal panacea for metabolic syndrome and type 2 diabetes? J Clin Endocrinol Metab 95: 4220–4222.

Osthoff G, Hugo A, de Wit M. 2007a. Milk composition of free-ranging springbok (*Antidorcas marsupialis*). Comp Biochem Phys B 146:421–426.

Osthoff G, Hugo A, de Wit M. 2007b. Milk composition of free-ranging sable antelope (*Hippotragus niger*). Mamm Biol 72:116–122.

Osthoff G, Hugo A, de Wit M. 2009a. Comparison of the milk composition of free-ranging blesbok, black wildebeest and blue wildebeest of the subfamily Alcelaphinae (family: Bovidae). Comp Biochem Phys B 154:48–54.

Osthoff G, Hugo A, de Wit M, Nguyen TPM. 2009b. The chemical composition of milk from free-ranging African buffalo (*Syncerus caffer*) S Afr J Wildl Res 39: 97–102.

Osthoff G, Hugo A, de Wit M. 2012. Comparison of the milk composition of free-ranging eland, kudu, gemsbok and scimitar oryx, with observations on lechwe, okapi and southern pudu. S Afr J Wildl Res 42: 23–34.

Ott I, Scott J. 1910. The action of infundibulin upon the mammary secretion. P Soc Exp Biol Med 8: 48–49.

Ozkan H, Tuzun F, Kumral A, Duman A. 2012. Milk kinship hypothesis in light of epigenetic knowledge. Clin Epigenet 4: 14.

Packer C, Lewis S, Pussey A. 1992. A comparative analysis of non-offspring nursing. Anim Behav 43: 265–281.

Parker H. 1956. Viviparous caecilians and amphibians. Nature 178: 250–252.

Parker K, Buckmaster C, Schatzberg A, Lyons D. 2005. Intranasal oxytocin administration attenuates the ACTH stress response in monkeys. Psychoneuroendocrinol 30: 924–929.

Partridge C, Shardo J, Boettcher A. 2007. Osmoregulatory role of the brood pouch in the euryhaline Gulf pipefish, Syngnathus scovelli. Comp Biochem Phys A 147(2): 556–561.

Perez PF, Dore J, Leclerc M, Levenez F, Benyacoub J, Serrant P, Segura-Roggero I, Schiffrin EJ, Donnet-Hughes A. 2007. Bacterial imprinting of the neonatal immune system: lessons from maternal cells. Pediatrics 119:e724–732.

Perry G, Dominy N, Claw K, Lee A, Fiegler H, Redon R, Werner J, Villanea F, Mountain J, Misra R, et al. 2007. Diet and the evolution of human amylase gene copy number variation. Nat Gen 39: 1256–1260.

Perry JC, Roitberg BD. 2005. Ladybird mothers mitigate offspring starvation risk by laying trophic eggs. Behav Ecol Sociobiol 58(6): 578–586.

Petraglia F, Florio P, Vale W. 2005. Placental expression of neurohormones and other neuroactive molecules in human pregnancy. In Birth, Distress and Disease: Placental-Brain

Interactions, Power ML, Schulkin J (eds.). Cambridge: Cambridge University Press, pp. 16-73.

Petzinger C, Oftedal OT, Jacobsen K, Murtough KL, Irlbeck NA, Power ML. 2014. Proximate composition of milk of the bongo (*Tragelaphus eurycerus*) in comparison to other African bovids and to hand-rearing formulas. Zoo Biol 33: 305-313.

Pfennig DW. 1992. Proximate and functional causes of polyphenism in an anuran tadpole. Funct Ecol 6: 167-174.

Pfennig DW, Reeve HK, Sherman PW. 1993. Kin recognition and cannibalism in spadefoot toad tadpoles. An Behav 46: 87-94.

Pieters B, Arntz O, Bennink M, Broeren M, van Caam A, Koenders M, van Lent P, van den Berg W, Vries M, van der Kraan P, et al. 2015. Commercial cow milk contains physically stable extracellular vesicles expressing immunoregulatory TGF-B. PLOS One 10: e0121123.

Pisanu S, Ghisaura S, Pagnozzi D, Biosa G, Tanca A, Roggio T, Uzzau S, Addis M. 2011. The sheep fat globule membrane proteome. J Proteomics 74: 350-358.

Place AR, Stoyan NC, Ricklefs RE, Butler RG. 1989. Physiological basis of stomach oil formation in Leach's storm-petrel (*Oceanodroma leucorhoa*). Auk 106: 687-699.

Pompei A, Cordisco L, Amaretti A, Zanoni S, Matteuzzi D, Rossi M. 2007. Folate production by bifidobacteria as a potential probiotic property. Appl Environ Microbiol 73: 179-185.

Posner A, Betts F. 1975. Synthetic amorphous calcium phosphate and its relation to bone mineral structure. Acc Chem Res 8: 273-281.

Poutahidis T, Kearney S, Levkovich T, Qi P, Varian B, Lakritz J, Ibrahim Y, Chatzigiagkos A, Alm E, Erdman S. 2013. Microbial symbionts accelerate wound healing via the neuro peptide hormone oxytocin. PLOS One 10: e78898.

Powe C, Pupolo K, Newburg D, lonnerdal B, Chen C, Allen M, Merewood A, Worden S, Welt C. 2011. Effects of recombinant human prolactin on breast milk composition. Pediatrics 127: e359-366.

Power ML, Oftedal O, Tardif S. 2002. Does the milk of callitrichid monkeys differ from that of larger anthropoids? Am J Primatol 56: 117-127.

Power ML, Ross CN, Schulkin J, Tardif SD. 2012. The development of obesity begins at an early age in captive common marmosets (*Callithrix jacchus*). Am J Primatol 74: 261-269.

Power ML, Ross CN, Schulkin J, Ziegler TZ, Tardif SD. 2013. Metabolic consequences of the early onset of obesity in common marmoset monkeys. Obesity 21: E592-598.

Power ML, Schulkin J, Drought H, Milligan LA, Murtough KL, Bernstein RM. Under review. Patterns of macronutrients and several bioactive molecules across lactation in a western lowland gorilla (*Gorilla gorilla*) and a Sumatran orangutan (*Pongo abelii*). Am J Primatol.

Power ML, Schulkin J. 2009. Evolution of Obesity. Baltimore: Johns Hopkins University Press.

Power ML, Schulkin J. 2012. The Evolution of the Human Placenta. Baltimore: Johns Hopkins University Press.

Power ML, Schulkin J. 2013. Maternal regulation of offspring development in mammals is an ancient adaptation tied to lactation. Appl Translational Genomics 2: 55-63.

Power ML, Verona CE, Ruiz-Miranda C, Oftedal OT. 2008. The composition of milk from free-living common marmosets (*Callithrix jacchus*) in Brazil. Am J Primatol 70: 78-83.

Powers GF. 1933. The alleged correlation between the rate of growth of the suckling and the composition of the milk of the species. J Pediatr 3: 201-217.

Prager E, Wilson A. 1988. Ancient origin of lactalbumin from lysozyme: analysis of DNA and amino acid sequences. J Mol Evol 27: 326-335.

Prentice A. 2000. Maternal calcium metabolism and bone mineral status. Am J Clin Nutr71: 1312s–1316s.

Queenan JT. 2011. Academy of Breastfeeding Medicine Founder's Lecture 2010: Breastfeeding: An obstetrician's view. Breastfeed Med 6: 7–14.

Ralls K, Kranz K, Lundrigan B. 1986. Mother-infant relationships in captive ungulates: variability and clustering. Anim Behav 34: 134–145.

Rambaran R, Serpell L. 2008. Amyloid fibrils: abnormal protein assembly. Prion 2: 112–117.

Ramos D. 2012. Breastfeeding: a bridge to addressing disparities in obesity and health. Breastfeed Med 7: 354–357.

Ranciaro A, Campbell M, Hirbo J, Ko W, Froment A, Anagnostou P, Kotze M, Ibrahim M, Nyambo T, Omar S, et al. 2014. Genetic origins of lactase persistance and the spread of pastoralism in Africa. Am J HumGenet 94: 496–510.

Rasmussen K. 2007. Association of maternal obesity before conception with poor lactation performance. Annu Rev Nutr 27: 103–121.

Rasmussen K, Kjolhede C. 2004. Prepregnant overweight and obesity diminish the prolactin response to suckling in the first week postpartum. Pediatrics 113: e465–471.

Raubenheimer D, Simpson S, Mayntz D. 2009. Nutrition, ecology and nutritional ecology: toward an integrated framework. Funct Ecol 23: 4–16.

Rautava S, Nanthakumar N, Dubert-Ferrandon A, Lu L, Rautava J, Walker W. 2011. Breast milk–transforming growth factor-β_2 specifically attenuates IL-1β-induced inflammatory responses in the immature human intestine via an SMAD6- and ERK-dependent mechanism. Neonatol 99: 192–201.

Rautava S, Walker W. 2009. Breastfeeding—an extrauterine link between mother and child. Breastfeed Med 4: 3–10.

Reich C, Arnould J. 2007. Evolution of Pinnipedia lactation strategies: a potential role for alpha- lactalbumin? Biol Lett 3: 546–549.

Reid L, Hubbell C. 1994. An assessment of the addiction potential of the opioid associated with milk. J Dairy Sci 77: 672–675.

Reifsnyder P, Churchill G, Leiter E. 2000. Maternal environment and genotype interact to establish diabesity in mice. Genome Res 10: 1568–1578.

Reik W, Lewis A. 2005. Co-evolution of X-chromosome inactivation and imprinting in mammals. Nature Reviews Genetics 6: 403–410.

Reinhart B, Slack F, Basson M, Pasquinelli A, Bettinger J, Rougvie A, Horvitz H, Ruvkun G. 2000. The 21-nucleotide let-7 RNA regulates developmental timing in *Caenorhabditis elegans*. Nature 403: 901–906.

Ren D, Lu G, Moriyama H, Mustoe A, Harrison E, French J. 2015. Genetic diversity in oxytocin ligands and receptors in New World Monkeys. PLOS One 10: 1–9.

Renfree M, Suzuki S, Kaneko-Ishino T. 2013. The origin and evolution of genomic imprinting and viviparity in mammals. PhilosT Roy Soc B 368: 2012.0151.

Rescigno M, Urbano M, Valzasina B, Francolini M, Rotta G, Bonasio R, Granucci F, Kraehenbuhl JP, Ricciardi-Castagnoli P. 2001. Dendritic cells express tight junction proteins and penetrate gut epithelial monolayers to sample bacteria. Nat Immun 2: 361–367.

Riddle O, Bates RW, Dykshorn SW. 1933. The preparation, identification and assay of prolactin—a hormone of the anterior pituitary. Am J Physiol 105: 191–216.

Riddle O, Braucher P. 1931. Studies on the physiology of reproduction in birds. XXX: Control of the special secretion of the crop-gland in pigeons by an anterior pituitary hormone. Am J Physiol 97: 617–625.

Ringel-Kulka T , Jensen E, McLaurin S, Woods E, Kotch JB, Labbok M, Bowling, JM, Dardess P, Baker S. (2011). Community-based participatory research of breastfeeding disparities in African American women. Infant Child Adolesc Nutr 3: 233–239.

Roberts J, Lillywhite H. 1980. Liquid barrier to water exchange in the reptile epidermis. Science 207: 1077–1079.

Roberts R, Jenkins K, Lawler T, Wegner F, Norcross J, Bernhards D, Newman J. 2001. Prolactin levels are elevated after infant carrying in parentally inexperienced common marmosets. Physiol Behav 72: 713–720.

Roff DA. 1993.(ed) Evolution of Life Histories: Theory and Analysis. Berlin: Springer Science and Business Media.

Roff DA. 2007. Contributions of genomics to life-history theory. Nature Reviews Genetics 8: 116–125.

Rood JP. 1980. Mating relationships and breeding suppression in the dwarf mongoose. Anim Behav 28: 143–150.

Rose-John S. 2012. IL-6 Trans-Signaling via the soluble IL-6 receptor: importance for the pro-inflammatory activities of IL-6. Int J Biol Sci 8: 1237–1247.

Ross CN, Power ML, Tardif SD. 2013. Relation of food intake behaviors and obesity development in young common marmoset monkeys. Obesity 21: 1891–1899.

Roth DE, Shah MR, Black RE, Baqui AH. 2010. Vitamin D status of infants in northeastern rural Bangladesh: preliminary observations and a review of potential determinants. J Health Popul Nutr 28: 458–469.

Russell J, Leng G. 1998. Sex, parturition and motherhood without oxytocin? J Endocrinol 157: 343–359.

Sackmann-Sala L, Guidotti J, Goffin V. 2015. Minireview: Prolactin regulation of adult stem cells. Mol Endocrinol 29: 667–681.

Sadler S. 1909. Infant Feeding by Artificial Means. London: George Routledge and Sons, Ltd.

Sagebakken G, Ahnesjo I, Mobley K, Goncalves I, Kvarnemo C. 2010. Brooding fathers, not siblings, take up nutrients from embryos. Proc R Soc Lond B Biol Sci 277: 971–977.

Samuelson L, Phillips R, Swanberg L. 1996. Amylase gene structures in primates: retroposon insertions and promoter evolution. Mol Biol Evol 13(6): 767–779.

Sanford LD, Nassar P, Ross RJ, Schulkin J, Morrison AR .1998. Prolactin microinjections into the amygdalar central nucleus lead to decreased NREM sleep. Sleep Res Online 1: 109–113.

Savino F, Liguori S, Fissore M, Oggero R. 2009. Breast milk hormones and their protective effect on obesity. Int J Ped Endocrinol, doi:10.1155/2009/327505.

Savino F, Liguori S, Sorrenti M, Fissore M, Oggero R. 2011. Breast milk hormones and regulation of glucose homeostasis. Int J Ped Endocrinol, doi:10.1155/2011/803985.

Scantlebury M, Russel AF, McIlrath GM, Speakman JR, Clutton-Brock TH. 2002. The nergetics of lactation in cooperatively breeding meerkats *Suricatta suricatta*. Proc R Soc Lond 269: 2147–2153.

Schafer E, Mackenzie K. 1911. The action of animal extracts on milk secretion. Proc R Soc Lond B Biol Sci 84: 16–22.

Scheele D, Wille A, Kendrick K, Stoffel-Wagner B, Becker B, Gunturkun O, Maier W, Hurlemann R. 2013. Oxytocin enhances brain reward system responses in men viewing the face of their female partner. PNAS 110: 20308–20313.

Schleithoff SS, Zittermann A, Tenderich G, Berthold HK, Stehle P, Koerfer R. 2006. Vitamin D supplementation improves cytokine profiles in patients with congestive heart failure: a double-blind, randomized, placebo-controlled trial. Am J Clin Nutr 83:754–759.Schneider J. 1996. Differential mortality and relative maternal investment in different life stages in *Stogodyphus lineatus* (Araneae, Erisdae). J Arachnology 24: 148–154.

Schneider J, Lubin Y. 1997. Infanticide by males in a spider with suicidal maternal care, *Stegodyphus lineatus* (Eresidae). AnBehav 54: 305–312.

Schoch R. 2009. Evolution of life cycles in early amphibians. Ann Rev Earth Planet Sci 37: 135–162.

Schroeder M, Shbiro L, Moran T, Weller A. 2010. Maternal environmental contribution to adult sensitivity and resistance to obesity in Long Evans rats. PLOS One 5(11): e13825.

Schultz M, Göttl C, Young R, Iwen P, and Vanderhoof J. 2004. Administration of oral probiotic bacteria to pregnant women causes temporary infantile colonization. J Pediatr Gastroenterol Nutr 38: 293–297.

Schwabl H. 1996. Environment modifies the testosterone levels of a female bird and its eggs. J Exp Zool 276: 157–163.

Schwartz S, Friedberg I, Ivanov I, Davidson L, Goldsby J, Dahl D, Herman D, Wang M, Donovan S, Chapkin R. 2012. A metagenomic study of diet-dependent interaction between gut microbiota and host in infants reveals differences in immune response. Genome Biol 13: r32.

Sela D, Mills D. 2010. Nursing our microbiota: molecular linkages between bifidobacteria and milk oligosaccharides. Trends Microbiol 18: 298–307.

Semba RD, Kumwenda N, Taha TE, Hoover DR, Lan Y, Eisinger W, Mtimavalye L, Broadhead R, Miotti PG, Van Der Hoeven L, et al. 1999. Mastitis and immunological factors in breast milk of lactating women in Malawi. Clin and DiagLab Immunol 6: 671–674.

Shackelton M, Vaillant F, Simpson K, Stingl J, Smyth G, Labat M, Wu L, Lindeman G, Visvader J. 2006. Generation of a functional mammary gland from a single stem cell. Nature 439: 84–88.

Shaikh U, Ahmed O. 2006. Islam and infant feeding. Breastfeeding Med 1: 164–167.

Sharp J, Lefevre C, Nicohlas K. 2007. Molecular evolution of monotreme and marsupial whey acidic protein genes. Evol and Devel 9: 378–392.

Shetty S, Bharathi L, Shenoy K, Hedge S. 1992. Biochemical properties of pigeon milk and its effects on growth. J Comp Physiol B 162: 632–636.

Shkolnik A, Maltz E, Choshniak I. 1980. The role of the ruminant's digestive tract as a water reservoir. In Digestive Physiology and Metabolism in Ruminants, Ruckebusch Y, Thivend P (eds.). Lancaster, England: MTP Press, pp. 731–742.

Shulman S, Friedmann H, Sims R. 2007. Theodor Escherich: the first pediatric infectious disease physician. CID 34: 1025–1029.

Singh GK, Kogan MD, Dee DL. 2007. Nativity/immigrant status, race/ethnicity, and socioeconomic determinants of breastfeeding initiation and duration in the United States, 2003. Pediatrics 119: S38–46.

Singhal A, Cole T, Fewtrell M, Kennedy K, Stephenson T, Elias-Jones A, Lucas A. 2007. Promotion of faster weight gain in infants born small for gestational age: is there an adverse effect on later blood pressure? Circulation 115: 213–220.

Singhal A, Kennedy K, Lanigan J, Clough H, Jenkins W, Elias-Jones A, Stephenson T, Dudek P, Lucas A. 2010. Dietary nucelotides and early growth in formula-fed infants: a randomized controlled trial. Pediatrics 126: e946–953.

Smith CW, Goldman AS. 1968. The cells of human colostrum. I. In vitro studies of morphology and function. Pediatr Res 2: 103–109.

Smith, H. 1772. Letters to Married Women on Nursing and the Management of Children. London.

Smith MI, Yatsunenko T, Manary MJ, Trehan I, Mkakosya R, Cheng J, Kau AL, Rich SS, Concannon P, Mychaleckyj JC, et al. 2013. Gut microbes of Malawian twin pairs discordant for kwashiorkor. Science 339: 548–554.

Smith V. 2011. Phylogeny of whey acidic protein (WAP) four-disulfide core proteins and their role in lower vertebrates and invertebrates. Biochem Soc Trans 39: 1403–1408.

Smithers R. 1983. The mammals of the Southern African subregion. Pretoria, South Africa: University of Pretoria.

Sowers M, Randolph J, Shapiro B, Jannausch M. 1995. A prospective study of bone density and pregnancy after an extended period of lactation with bone loss. Obstet Gynecol 85: 285–289.

Spangenberg J, Jacomet S, Schibler J. 2006. Chemical analyses of organic residues in archaeological pottery from Arbon Bleiche 3, Switzerland: evidence for dairying in the late Neolithic. J Archaeolog Sci 33: 1–13.

Sparks P. 2011. Racial/ethnic differences in breastfeeding duration among WIC-eligible families. Women's Health Issues 21: 374–382.

Stavropoulos K, Carver L. 2013. Research Review: Social motivation and oxytocin in autism-implications for joint attention development and intervention. J Child Psychol Psychiatry 54: 603–618.

Stearns SC. 1992. The Evolution of Life Histories. Oxford: Oxford University Press.

Stearns SC, Allal N, Mace R. 2008. Life history theory and human development. In Foundations of Evolutionary Psychology. Crawford C, Krebs D (eds.). New York: Taylor and Francis Group/Lawrence Erlbaum Associates, pp. 47–69.

Stenchever M, Hale R, Rowe N. 2007. Special Report from ACOG. Breastfeeding: maternal and infant aspects. ACOG Clin Rev 12: 1S–16S.

Stevens E, Patrick T, Pickler R. 2009. A history of infant feeding. J Perinatal Ed 18: 32–39.

Størdal K, White R, Eggesbo M. 2013. Early feeding and risk of celiac disease in a prospective birth cohort. Pediatrics 132: 1202–1209.

Strasser B, Mlitz V, Hermann M, Tschachler E, Eckhart L. 2015. Convergent evolution of cysteine-rich proteins in feathers and hair. BMC Evol Biol. 15: 82.

Stricker P, Grueter R. 1928. Action of the anterior lobe of the pituitary gland on milk secretion. C R Biol 99: 1978–1980

Stubbs J, Lekutis C, Singer K, Bui A, Yuzuki D, Srinivasan U, Parry G. 1990. cDNA cloning of a mouse mammary epithelial cell surface protein reveals the existence of epidermal growth factor-like domains linked to factor VIII-like sequences. Proc Natl Acad Sci 87: 8417–8421.

Sullivan S, Schanler RJ, Kim JH, Patel AL, Trawöger R, Kiechl-Kohlendorfer U, Chan GM, Blanco CL, Abrams S, Cotten CM, et al. 2010. An exclusively human milk-based diet is associated with a lower rate of necrotizing enterocolitis than a diet of human milk and bovine milk-based products. J Pediatr 156: 562–567.

Suzuki YA, Shin K, Lönnerdal B. 2001 Molecular cloning and functional expression of a human intestinal lactoferrin receptor. Biochemistry 40: 15771–15779.

Swaminathan N. 2007. Strange but true: males can lactate. Sci Am. http://www.scientificamerican.com/article/strange-but-true-males-can-lactate/.

Tabor N. 2013. Wastelands of tropical Pangea: high heat in the Permian. Geology 41: 623–624.

Takahashi T, Hayashi K, Hosoe M. 2013. Biology of the placental proteins in domestic ruminants: expression, proposed roles and practical applications. Japan Agric Res Quart 47: 43–45.

Takuwa-Kuroda K, Iwakoshi-Ukena E, Kanda A, Minakata H. 2003. Octopus, which owns the most advanced brain in invertebrates, has two members of vasopressin/oxytocin superfamily as in vertebrates. Regul Peptides 115(2): 139–149.

Tao N, Wu S, Kim J, An H, Hinde K, Power M, Gagneux P, German J, Lebrilla C. 2011. Evolutionary glycomics: characterization of milk oligosaccharides in primates. J Proteome Res 10: 1548–1557.

Tardif SD, Power M, Oftedal OT, Power RA, Layne DG. 2001. Lactation, maternal behavior and infant growth in common marmoset monkeys (*Callithrix jacchus*): effects of maternal size and litter size. Behav Ecol Sociobiol 51: 17–25.

Taylor BA, Varga GA, Whitsel TJ, Hershberger TV. 1990. Composition of blue duiker (*Cephalophus monticola*) milk and milk intake by the calf. Small Rum Res 3: 551–560.

Thome M, Alder E, Ramel A. 2006. A population-based study of exclusive breastfeeding in Icelandic women: is there a relationship with depressive symptoms and parenting stress? Int J Nursing Studies 43: 11–20.

Tilden C, Oftedal O. 1997. Milk composition reflects pattern of maternal care in prosimian primates. Am J Primatol 41: 195–211.

Tishkoff SA, Reed FA, Ranciaro A, Voight BF, Babbitt CC, Silverman JS, Deloukas P. 2007. Convergent adaptation of human lactase persistence in Africa and Europe. Nat Gen 39: 31–40.

Title, A., Denzler, R., Stoffel, M. 2015. Uptake and function studies of maternal milk–derived microRNAs. J Biol Chem 290: 23680–23691.

Tizo-Pedroso E, Del-Claro K. 2005. Matriphagy in the neotropical pseudoscorpion *Paratemnoids Nidificator* (Balzan 1888) (Atemindae). J Arachnology 33: 873–877.

Tizo-Pedroso E, Del-Claro K. 2007. Cooperation in the neotropical pseudoscorpion, *Paratemnoides nidificator* (Balzan, 1888): feeding and dispersal behavior. Insect Soc 54: 124–131.

Trivers R. 1972. Parental investment and sexual selection. In Sexual Selection and the Descent of Man, 1871–1971, Campbell B (ed.). Chicago: Aldine, pp. 136–179.

Trott JF, Simpson KJ, Moyle RLC, Hearn CM, Shaw G, Nicholas KR, Renfree MB. 2003. Maternal regulation of milk composition, milk production, and pouch young development during lactation in the Tammar wallaby (*Macropus eugenii*). Biol Reprod 68: 929–936.

Truswell A. 2005. The A2 milk case: a critical review. Eur J Clin Nutr 59: 623–631.

Tsvetov G, Levy S, Benbassat C, Shraga-Slutzky I, Hirsch D. 2014. Influence of number of deliveries and total breast-feeding time on bone mineral density in premenopausal and young postmenopausal women. Maturitas 77: 249–254.

Tzotzas T, Papadopoulou F, Tziomals K, Karras S, Gastaris K, Perros P, Krassas G. 2010. Rising serum 25-hydroxy-vitamin D levels after weight loss in obese women correlate with improvement in insulin resistance. J Clin Endocrinol Metab 95: 4251–4257.

Underwood MA, Gilbert WM, Sherman MP. 2005. Amniotic fluid: not just fetal urine anymore. J Perinatol 25: 341–348.

Urashima T, Fukuda K, Messer M. 2012. Evolution of milk oligosaccharides and lactose: a hypothesis. Animal 6: 369–374.

U.S. Department of Agriculture Food and Nutrition Service. 2009. The WIC program. Retrieved from: http://www.fns.usda.gov/wic.

Uvnas-Moberg K. 1998. Oxytocin may mediate the benefits of positive social interaction and emotions. Psychoneuroendocrinol23: 819–835.

Vandebergh W, Bossuyt F. 2012. Radiation and functional diversification of alpha keratins during early vertebrate evolution. Mol Biol Evol 29: 995–1004.

Van de Perre P. 2003. Transfer of antibody via mother's milk. Vaccine 21: 3374–3376.

van De Sande M, van Buul V, Brouns F. 2014. Autism and nutrition: the role of the gut-brain axis. NutrRes Rev 27: 199–214.

Visvader JE, Stingl J. 2014. Mammary stem cells and the differentiation hierarchy: current status and perspectives. Genes Develop 28: 1143–1158.

Vorbach C, Capecchi M, Penninger J. 2006. Evolution of the mammary gland from the innate immune system. BioEssays 28: 606–616.

Wagner C. 2002. Amniotic fluid and human milk: a continuum of effect? J Pediatr Gàstroenterol Nutr 34: 513–514.

Wagner C, Greer F. 2008. Prevention of rickets and vitamin D deficiency in infants, children, and adolescents. Pediatrics 122: 1142–1152.

Wagner C, Taylor S, Johnson D. 2008. Host factors in amniotic fluid and breast milk that contribute to gut maturation. Clin Rev Allerg Immunol 34: 191–204.

Wall C, Grant C, Jones I. 2013. Vitamin D status of exclusively breastfed infants aged 2–3 months. Arch Dis Child 98: 176–179.

Wallis M. 2009. Prolactin in the Afrotheria: characterization of genes encoding prolactin in elephant (*Loxodonta africana*), hyrax (*Procavia capensis*) and tenrec (*Echinops telfairi*). J Endocrinol 200: 233–240.

Wallis M. 2012. Molecular evolution of the neurohypophysial hormone precursors in mammals: comparative genomics reveals novel mammalian oxytocin and vasopressin analogues. Gen Comp Endocrinol179: 313–318.

Ward L, Gaboury I, Ladhani M, Zlotkin S. 2007. Vitamin D-deficiency rickets among children in Canada. CMAJ 177: 161–166.

Ward R, Ninonuevo M, Mills D, Lebrilla C, German J. 2006. In vitro fermentation of breast milk oligosaccharides by *Bifidobacterium infantis* and *Lactobacillus gasseri*. Appl Environ Microbiol 72: 4497–4499.

Watson JD, Crick FH. 1953. Molecular structure of nucleic acids. Nature 171: 737–738.

Weber JA, Baxter DH, Zhang S, Huang DY, Huang KH, Lee MJ, Galas DJ, Wang K. H, 2010. The microRNA spectrum in 12 body fluids. Clin Chem 56: 1733–1741.

Wehr T, Moul D, Barbato G, Giesen H, Seidel J, Barker C, Bender C. 1993. Conservation of photoperiod-responsive mechanisms in humans. Am J Physiol265: R846–57.

Weyer C, Funahashi T, Tanaka S, Hotta K, Matsuzawa Y, Pratley R, Tataranni P. 2001. Hypoadiponectinemia in obesity and type 2 diabetes: close association with insulin resistance and hyperinsulinemia. J Clin Endocrinol Metab86: 1930–1935.

Wickes I. 1953. A history of infant feeding. Arch Dis Child 28: 151–158

Widdowson E, Dickerson J. 1960. The effect of growth and function on the chemical composition of soft tissues. Biochem J 77: 30.

Wiebe K. 1996. The insurance-egg hypothesis and extra reproductive value of last-laid eggs in clutches of American Kestrels. Auk 113: 258–261.

Wiley AS. 2012. Cow milk consumption, insulin-like growth factor-I, and human biology: a life history approach. Am J Hum Biol, 24: 130–138.

Wilkinson GS. 1992. Communal nursing in the evening bat, *Nycticeius humeralis*. Behav Ecol Sociobiol 31: 225–235.

Williams L, Gibson S, McDaniel M, Bazzel J, Barnes S, Abee C. 1994. Allomaternal interactions in the Bolivian squirrel monkey (*Saimiri boliviensis boliviensis*). Am J Primatol 34: 145–156.

Wilson DE, Hirst SM. 1977. Ecology and factors limiting roan and sable antelope populations in South Africa. Wildlife Monogr 54:1–111.

Winslow J, Insel T. 2002. The social deficits of the oxytocin knockout mice. Neuropeptides 36: 221–229.

Woese C, Fox G. 1977. Phylogenetic structure of the prokaryotic domain: the primary kingdoms. Proc Natl Acad Sci USA 74: 5088–5090.

Wolf M, Van Doorn GS, Leimar O, Weissing FJ. 2007. Life-history trade-offs favour the evolution of animal personalities. Nature 447: 581–584.

Wongdee K, Charoenphandhu N. 2012. Regulation of epithelial calcium transport by prolactin: from fish to mammals. Gen Comp Endocrinol 181: 235–240.

Woo JG, Guerrero ML, Altaye M, Ruiz-Palacios GM, Martin LJ, Dubert-Ferrandon A, Newburg DS, Morrow AL. 2009. Human milk adiponectin is associated with infant growth in two independent cohorts. Breastfeed Med 4: 101–109.

Woo JG, Guerrero ML, Guo F, Martin LJ, Davidson BS, Ortega H, Ruiz-Palacios GM, Morrow AL. 2012. Human milk adiponectin affects infant weight trajectory during the second year of life. J Pediatr Gastroenterol Nutr 54: 532–539.

Wright AL, Holberg CJ, Taussig LM, Martinez FD. 2001. Factors influencing the relation of infant feeding to asthma and recurrent wheeze in childhood. Thorax 56: 192–197.

Xi D, Peng Y G, Ramsdell J. S. 1997. Domoic acid is a potent neurotoxin to neonatal rats. Nat Toxins 5: 74–79.

Yamshchikov A, Desai N, Blumberg H, Ziegler T, Tangpricha V. 2009. Vitamin D for treatment and prevention of infectious diseases: a systematic review of randomized controlled trials. Endocr Practice 15: 438–449.

Young W, Shepard E, Emico J, Hennighausen L, Wagner K, LaMarca M, McKinney, Gins E. 1996. Deficiency in mouse oxytocin prevents milk ejection, but not fertility or parturition. J Neuroendocrinol 8: 847–853.

Zhang R, Naughton D. 2010. Vitamin D in health and disease: current perspectives. NutrJ 9: 65.

Zhou Q, Mingzhou L, Wang X, Li Q, Wang T, Zhu Q, Zhou X, Wang X, Gao X, Li X. 2012. Immune-related microRNAs are abundant in breast milk exosomes. Int J Biol Sci 8: 118–123.

Ziegler E, Hollis B, Nelson S, Jeter J. 2006. Vitamin D deficiency in breastfed infants in Iowa. Pediatrics 118: 603–610.

Ziemke F, Mantzoros C. 2010. Adiponectin in insulin resistance: lessons from translational research. AmJ ClinNutr 91: 258S–261S.

Zinger M, McFarland M, Ben-Jonathan N. 2003. Prolactin expression and secretion by human breast glandular and adipose tissue explants. J ClinEndocrinol Metab 88: 689–696.

Zittermann A, Iodice S, Pilz S, Grant W, Bagnardi V, Gandini S. 2012. Vitamin D deficiency and mortality risk in general population: a meta-analysis of prospective cohort studies. Am J Clin Nutr 95: 91–100.

Index

Boldface entries refer to figures (**f**) and tables (**t**); *italic* entries refer to text boxes.